Springer Series in Statistics

Springer
New York
Berlin
Heidelberg
Hong Kong
London
Milan
Paris
Tokyo

Springer Series in Statistics

Andersen/Borgan/Gill/Keiding: Statistical Models Based on Counting Processes.
Atkinson/Riani: Robust Diagnostic Regression Analysis.
Berger: Statistical Decision Theory and Bayesian Analysis, 2nd edition.
Borg/Groenen: Modern Multidimensional Scaling: Theory and Applications
Brockwell/Davis: Time Series: Theory and Methods, 2nd edition.
Chan/Tong: Chaos: A Statistical Perspective.
Chen/Shao/Ibrahim: Monte Carlo Methods in Bayesian Computation.
David/Edwards: Annotated Readings in the History of Statistics.
Devroye/Lugosi: Combinatorial Methods in Density Estimation.
Efromovich: Nonparametric Curve Estimation: Methods, Theory, and Applications.
Eggermont/LaRiccia: Maximum Penalized Likelihood Estimation, Volume I: Density Estimation.
Fahrmeir/Tutz: Multivariate Statistical Modelling Based on Generalized Linear Models, 2nd edition.
Fan/Yao: Nonlinear Time Series: Nonparametric and Parametric Methods.
Farebrother: Fitting Linear Relationships: A History of the Calculus of Observations 1750-1900.
Federer: Statistical Design and Analysis for Intercropping Experiments, Volume I: Two Crops.
Federer: Statistical Design and Analysis for Intercropping Experiments, Volume II: Three or More Crops.
Ghosh/Ramamoorthi: Bayesian Nonparametrics.
Glaz/Naus/Wallenstein: Scan Statistics.
Good: Permutation Tests: A Practical Guide to Resampling Methods for Testing Hypotheses, 2nd edition.
Gouriéroux: ARCH Models and Financial Applications.
Gu: Smoothing Spline ANOVA Models.
Györfi/Kohler/Krzyżak/ Walk: A Distribution-Free Theory of Nonparametric Regression.
Haberman: Advanced Statistics, Volume I: Description of Populations.
Hall: The Bootstrap and Edgeworth Expansion.
Härdle: Smoothing Techniques: With Implementation in S.
Harrell: Regression Modeling Strategies: With Applications to Linear Models, Logistic Regression, and Survival Analysis
Hart: Nonparametric Smoothing and Lack-of-Fit Tests.
Hastie/Tibshirani/Friedman: The Elements of Statistical Learning: Data Mining, Inference, and Prediction
Hedayat/Sloane/Stufken: Orthogonal Arrays: Theory and Applications.
Heyde: Quasi-Likelihood and its Application: A General Approach to Optimal Parameter Estimation.
Huet/Bouvier/Gruet/Jolivet: Statistical Tools for Nonlinear Regression: A Practical Guide with S-PLUS and R Examples, 2nd edition.
Ibrahim/Chen/Sinha: Bayesian Survival Analysis.
Jolliffe: Principal Component Analysis.

(continued after index)

Thomas J. Santner
Brian J. Williams
William I. Notz

The Design and Analysis of Computer Experiments

Springer

Thomas J. Santner
Department of Statistics
Ohio State University
1958 Neil Avenue
Columbus, OH 43210
USA
tjs@stat.ohio-state.edu

Brian J. Williams
Department of Statistics
Ohio State University
1958 Neil Avenue
Columbus, OH 43210
USA
brianw@rand.org

William I. Notz
Department of Statistics
Ohio State University
1958 Neil Avenue
Columbus, OH 43210
USA
win@stat.ohio-state.edu

Library of Congress Cataloging-in-Publication Data
Santner, Thomas J., 1947–
Design & analysis of computer experiments / Thomas J. Santner, Brian J. Williams, William I. Notz.
p. cm. — (Springer series in statistics)
Includes bibliographical references and index.
ISBN 0-387-95420-1 (acid-free paper)
1. Experimental design. 2. Physical sciences—Experiments—Computer simulation. I. Title: Design and analysis of computer experiments. II. Williams, Brian J. III. Notz, William. IV. Title. V. Series.
QA279.S235 2003
519.5—dc21 2003045444

ISBN 0-387-95420-1 Printed on acid-free paper.

Printed in the United States of America.

9 8 7 6 5 4 3 2 1 SPIN 10863256

Typesetting: Pages created by the authors using a Springer LaTeX 2e macro package.

www.springer-ny.com

Springer-Verlag New York Berlin Heidelberg
A member of BertelsmannSpringer Science+Business Media GmbH

To Gail, Aparna, and Claudia

for their encouragement and patience

Contents

1
Physical Experiments and Computer Experiments

1.1 Introduction

This book describes methods for designing and analyzing research studies that are conducted using computer code in lieu of a physical experiment. Historically, Statistics has been the scientific discipline that creates methodology for conducting empirical research. The process of designing a study to answer a specific research question first addresses the problem of identifying the variables to be observed, i.e., what data to collect. Traditional methods of data collection include retrospective techniques such as cohort studies and the case-control studies used in epidemiology. The gold standard data collection method for establishing cause and effect relationships is the prospective designed experiment. *Agricultural field experiments* were one of the first subject matter disciplines that used designed experiments. Over time, many other subject matter areas and modes of experimentation have been developed. For example, *controlled clinical trials* are used extensively in studying medical therapies and *simulation experiments* are used extensively in operations research to compare the performance of (well) understood physical systems having stochastic components such as the flow of material through a job shop. Once the research study has been designed and executed, Statistics either identifies or develops appropriate methods to analyze the resulting data. This is true whether the data are generated by an actual physical process or by a computer code that is intended to simulate such a process. However, there are differences between data generated by a physical experiment and data generated by a computer code.

Physical experiments measure a stochastic response corresponding to a set of (experimenter determined) treatment input variables. Unfortunately, most experiments also involve nuisance input variables that may or may not be recognized and cause some of the variation in the experimental response. Faced with this reality, statisticians have developed a variety of techniques to increase the validity of treatment comparisons for physical experiments. One such method is *randomization*. Randomizing the order of applying the experimental treatments is done to prevent unrecognized nuisance variables from systematically affecting the response in such a way as to be confounded with treatment variables. Another technique to increase experimental validity is *blocking*. Blocking is used when there are recognized nuisance variables, such as different locations or time periods, for which the response is expected to behave differently, even in the absence of treatment variable effects. Yields from fields in northern climates can be expected to be different from those in southern climates and males may react differently to a certain medical therapy than females. A block is a group of experimental units that have been predetermined to be homogeneous. By applying the treatment variable in a special way to blocks, comparisons can be made among the units that are as similar as possible, except for the treatment of interest. *Replication* is a third technique for increasing the validity of an experiment. Adequate replication means that an experiment is run on a sufficiently large scale to prevent the unavoidable "measurement" variation in the response from obscuring treatment differences.

In the past 15 to 20 years, yet another mode of conducting experiments has become increasingly popular. Suppose that a mathematical theory exists, e.g., a set of differential equations, that relates the output of a complex physical process to a set of input variables. Suppose also that a numerical method exists for accurately solving the mathematical system. The presence of these two elements with appropriate computer hardware and software to implement the numerical methods allows one to conduct a *computer experiment* that produces a "response" corresponding to any given set of input variables. While many computer experiment codes are solutions to large systems of differential equations, other codes are simply extremely sophisticated simulations run to the point that the simulation error is essentially zero.

In some cases computer experimentation is feasible when physical experimentation is impossible. For example, the number of input variables may be too large to consider performing a physical experiment, there may be ethical reasons why a physical experiment cannot be run, or it may simply be economically prohibitive to run an experiment on the scale required to gather sufficient information to answer a particular research question (nevertheless, see Section 7.2 for cautions about situations in which a computer experiments are possible with minimal physical ground-truth—there is no free lunch here).

Examples of scientific and technological developments that have been conducted using computer codes are many and growing. They have been used to predict climate and weather, the performance of integrated circuits, the behavior of controlled nuclear fusion devices, the properties of thermal energy storage devices, and the stresses in prosthetic devices. More detailed motivating examples will be provided in Section 1.2.

In contrast to classical physical experiments, a computer experiment yields a deterministic answer for a given set of input conditions; indeed, the code produces *identical* answers if run twice using the same set of inputs. Thus, using randomization to avoid potential confounding of treatment variables due to run order is irrelevant. Similarly, blocking the runs into groups that represent runs on experimental units "more nearly alike" is also irrelevant. Indeed, none of the traditional principles of blocking, randomization, and replication are of use in solving the design and analysis problems associated with computer experiments. However, we still use the word "experiment" to describe such a code because the goal in both physical and computer experiments is to determine which treatment variables affect a given response and, for those that do, the details of the input-output relationship.

In addition to being deterministic, computer experiments can involve code that is time-consuming to run; in some finite element models, it would not be unusual for code to run for 12 hours or even considerably longer to produce a single response. A second feature of many computer experiments is that the number of input variables can be quite large–15 to 20 or more variables. One reason for the large number of variables is that the inputs must include not only engineering variables (variables that can be set by an engineer or scientist to control the process of interest) but also variables that represent operating conditions and/or variables that are part of the code and would be known in a perfect world. For example, the strain occurring at the bone–prosthesis boundary of a knee implant depends on the loading of the joint and on the material properties of the bone at the bone–prosthesis interface. These are patient specific. Other examples of problems with both engineering and other types of input variables are given in Section 2.1.

This book will discuss methods that can be used to design and analyze computer experiments that account for their special features. The remainder of this chapter will provide several motivating examples and an overview of the book.

1.2 Some Examples of Computer Models

This section sketches several areas where computer models are used. Our intent is to show the *breadth* of application of such models and to provide some feel for the types of inputs and outputs used by these models. The details of the mathematical models implemented by the computer code will not be given, but references to the source material are provided for interested readers.

1.2.1 Evolution of Fires in Enclosed Areas

Deterministic computer models are used in many areas of fire protection design including egress (exit) analysis. We describe one of the early "zone computer models" that is used to predict the fire conditions in an enclosed room. Cooper (1980) and Cooper and Stroup (1985) provided a mathematical model and its implementation in FORTRAN for describing the evolution of a fire in a single room with closed doors and windows that contains an object at some point below the ceiling that has been ignited. The room is assumed to contain a small leak at floor level to prevent the pressure from increasing in the room. The fire releases both energy and hot combustion by-products. The rate at which energy and the by-products are released is allowed to change with time. The by-products form a plume which rises towards the ceiling. As the plume rises, it draws in cool air, which decreases the plume's temperature and increases its volume flow rate. When the plume reaches the ceiling, it spreads out and forms a hot gas layer whose lower boundary descends with time. There is a relatively sharp interface between the hot upper layer and the air in the lower part of the room, which in this model is considered to be at air temperature. The only interchange between the air in the lower part of the room and the hot upper layer is through the plume. The model used by these programs can therefore be described as a two "zone" model.

The Cooper and Stroup (1985) code is called ASET (Available Safe Egress Time). Walton (1985) implemented their model in BASIC, calling his computer code ASET-B; he intended his program to be used in the first generation of personal computers available at that time of its development. ASET-B is a compact, easy to run program that solves the same differential equations as ASET using a simpler numerical technique.

The *inputs* to ASET-B are

- the room ceiling height and the room floor area,
- the height of the burning object (fire source) above the floor,
- a heat loss fraction for the room (which depends on the insulation in the room, for example),
- a material-specific heat release rate, and
- the maximum time for the simulation.

The program *outputs* are the *temperature* of the hot smoke layer and its *distance* above the fire source as a function of time.

Since these early efforts, computer codes have been written to model wildfire evolution as well as fires in confined spaces. As typical examples of this work, we point to Lynn (1997) and Cooper (1997), respectively. The publications of the Building and Fire Research Laboratory of NIST can be found online at `http://fire.nist.gov/bfrlpubs/`. Finally, we mention the review article by Berk, Bickel, Campbell, Fovell, Keller-McNulty, Kelly, Linn, Park, Perelson, Rouphail, Sacks and Schoenberg (2002) which describes statistical approaches for the evaluation of computer models for wildfires. Sahama and Diamond (2001) give a case study using the statistical methods introduced in Chapter 3 to analyze a set of 50 observations computed from the ASET-B model.

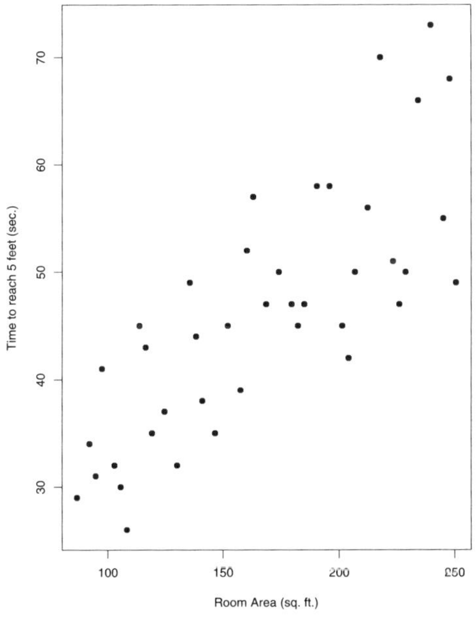

FIGURE 1.1. Scatterplot of room area versus the time for the hot smoke layer to reach five feet above the fire source.

To provide a sense of the effect of each of these variables on the evolution of the fire, we *fixed* the heat release rate to correspond to fire material that constitutes a "semi-universal" fire; this heat release profile corresponds to a fire in a "fuel package consisting of a polyurethane mattress with sheets and fuels similar to wood cribs and polyurethane on pallets and commodities in paper cartons stacked on pallets." (Birk (1997)). Then we varied the remaining four factors using a "Sobol´ design" (Sobol´ designs are described in Section 5.5). We computed the time, to the nearest second, for the fire to reach five feet above the burning fuel package, the fire source.

Scatterplots were constructed of each input versus the time required by hot smoke layer to reach five feet above the fire source. Only room area showed strong visual associations with the output; Figure 1.1 shows this scatterplot (see also Figure 3.10 for all four plots). This makes intuitive sense because more by-product is required to fill the top of a large room and hence, longer times are required until this layer reaches a point five feet above the fire source. The data from this example will be used later to illustrate several analysis methods.

1.2.2 Design of Prosthesis Devices

Biomechanical engineers have been using computer models to determine the performance of total joint replacements for more than two decades. These models supplement clinical studies and limited physical experiments. Computer models are ideal for design comparisons and optimization. Two difficulties with the use of many computer models are their long running times and their validation, an area of ongoing research (Section 7.2 discusses validation). For example, even using a fast workstation, a single run of many such codes can require hours or even days.

As a simple example, Chang, Williams, Bawa Bhalla, Belknap, Santner, Notz and Bartel (2001) consider the design of a "bullet" tip hip prosthesis (specifically, a cementless cobalt chrome Ranawat-Burstein implant with a 16 mm diameter, 100 mm long distal stem, and manufactured by Biomet, Inc., Warsaw, IN). Figure 1.2 is a line drawing that illustrates the prosthesis. The finite element code of Chang et al. (2001) calculated two *outputs*. The first was "femoral stress shielding," a measure of how much of the load is deflected from the upper medial femur (the section of the femur near the top of the prosthesis, directly under the hip ball); femoral stress shielding should be minimized to prevent bone loss at the neck of the prosthesis. The second output of the computer code was "implant toggling," a measure of the flexing of the implant in the coronal plane of the body; excessive flexing can cause the implant to loosen and thus implant toggling was also to be minimized.

The goal of their study was to consider alternatives to the current design in which two of the dimensions were allowed to vary. These engineering design variables were

- b, the length of a bullet tip, and
- d, the midstem diameter,

which are illustrated in Figure 1.2. In addition, the model had three patient-specific variables that also contributed to the structural response:

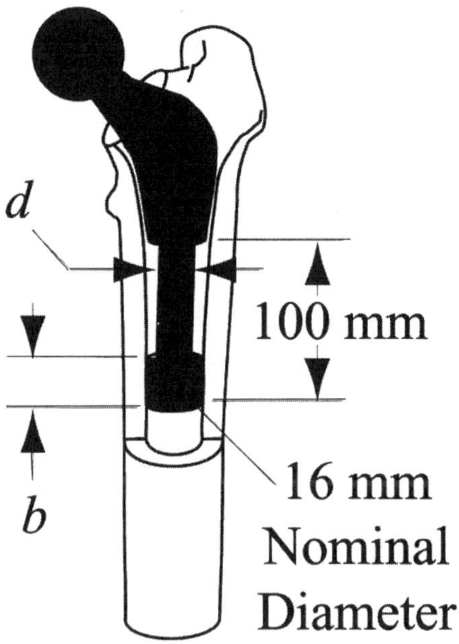

FIGURE 1.2. Control variables b and d for a class hip prosthesis.

- Θ, the joint force angle,
- E, the elastic modulus of the surrounding cancellous bone, and
- the implant-bone interface friction.

Figure 1.3 illustrates the variables (Θ, E). The (approximate) values of these three input variables in human populations were determined from studies in the orthopedic literature. For example, the joint force angle was determined in telemetric hip force studies (Kotzar, Davy, Berilla and Goldberg (1995)).

An interesting feature of Chang et al. (2001) is that the two outputs represented competing objectives. Flexible prostheses minimize stress shielding but permit the prosthesis toggling and thus increase the chance of loosening. Stiff prostheses will not toggle from side to side but will shield the bone at the neck of the prosthesis from experiencing stresses causing this section of the bone to be gradually resorbed by the body. The modified structure can be more fragile and can fracture under the stress of, say, a fall. Combining multiple objectives is typically problematic and Subsection 2.2.4 describes one method of doing so that has mathematical, and often practical, appeal.

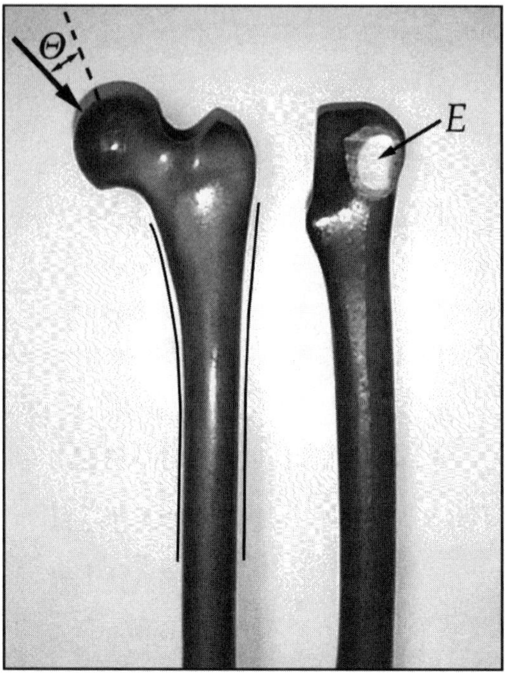

FIGURE 1.3. Environmental variables Θ and E.

1.2.3 Formation of Pockets in Sheet Metal

Montgomery and Truss (2001) discussed a computer model that determines the failure depth of symmetric rectangular pockets that are punched in automobile steel sheets; the failure depth is the depth at which the sheet metal tears. Sheet metal, suitably formed in this manner, is used to fabricate many parts of automobiles. This application is but one of many examples of computer models used in the automotive industry.

Rectangular pockets are formed in sheet metal by pressing the metal sheet with a punch which is a target shape into a conforming die. There are *six input variables* to the Montgomery and Truss (2001) code, all of which are engineering design variables. These variables can either be thought of as characteristics of the punch/die machine tool used to produce the pockets or, in most cases, as characteristics of the resulting pockets.

Five of the variables can easily be visualized in terms of the pocket geometry. In a top view of the pocket, Figure 1.4 illustrates the *length* l and the *width* w of the rectangular pocket (defined to omit the curved corner of the pocket). In a side view of the pocket, Figure 1.5 shows the *fillet radius* f, which is the radius of the circular path that the metal follows as it curves from the flat upper metal surface to the straight portion of the pocket wall; this region is denoted R in both the side and top views of the pocket. The same fillet radius is followed as the straight portion of the pocket wall

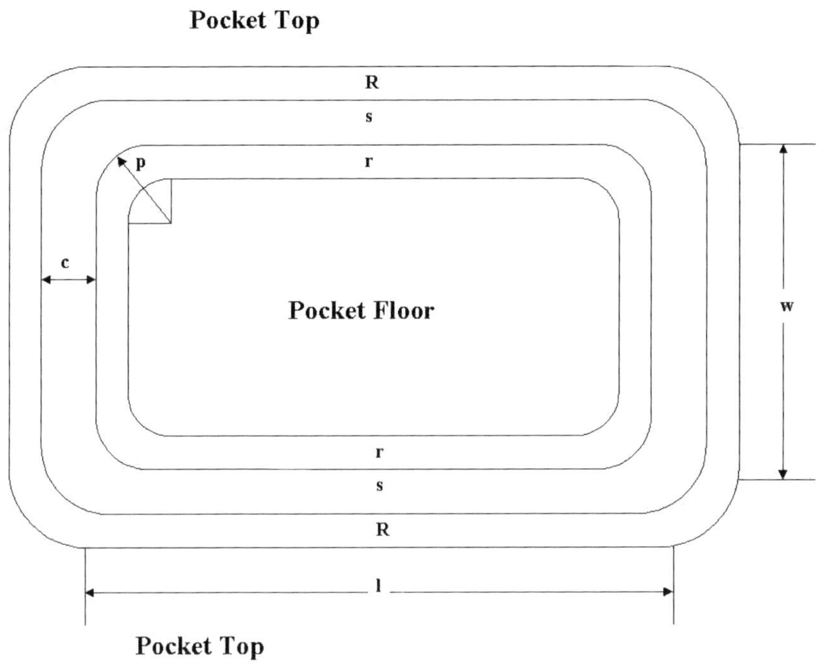

FIGURE 1.4. Top view of the pocket formed by a punch and die operation. The floor of pocket is the innermost rectangle. The regions R, s, and r correspond to the similarly labeled regions in the side view.

curves in a circular manner to the pocket floor; this region is denoted by r in both views. Viewed from the top in Figure 1.4, the *clearance* is the horizontal distance c during which the angled side wall descends vertically to the pocket floor in Figure 1.5. In terms of the punch/die manufacturing tool, the clearance is the distance between the punch and the die when the punch is moved to its maximum depth within the die; the distance between the two tool components is constant. Lastly, the *punch plan view radius* p is illustrated in Figure 1.4. The *lock bead distance*, shown in Figure 1.5, is a distance d measured away from the pocket edge on the top metal surface; the machine tool does not allow stretching of the sheet metal beyond the distance d from the pocket edge.

To provide a sense of the (marginal) effect of each of these variables on the failure depth, we plotted the failure depth versus each of the six explanatory variables for the set of 234 runs analyzed by Montgomery and Truss (2001). Two of these scatterplots are shown in Figure 1.6; they are representative of the six marginal scatterplots. Five variables are only weakly related to failure depth and the panel in Figure 1.6 showing failure depth versus fillet radius is typical of these cases. One variable, clearance, shows a strong relationship with failure depth.

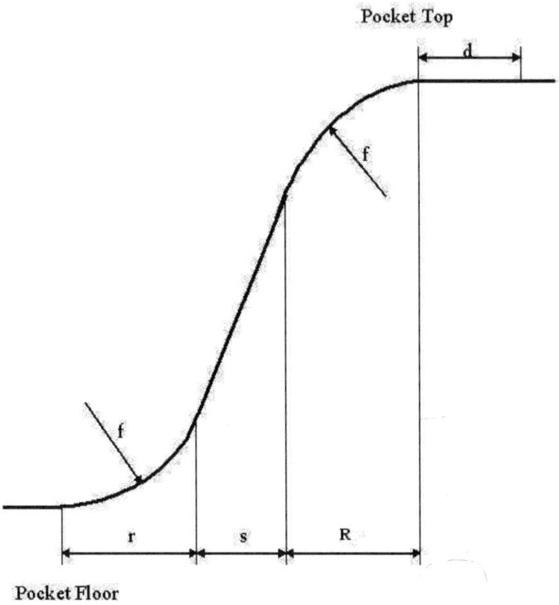

FIGURE 1.5. Side view of part of a symmetric pocket formed by a punch and die operation. The angled side wall is created by the same fillet angle at the top by the die and at the bottom by the edge of the punch.

1.2.4 Other Examples

The purpose of this subsection is to sketch several applications of computer models that involve *large* numbers of input variables compared with the models described in Subsections 1.2.1–1.2.3. These examples will also serve to broaden the reader's appreciation of the many scientific and engineering applications of such models. Finally, as an additional source of motivating examples, we again remind the reader of Berk et al. (2002), who report on a workshop that discussed computer models in four diverse areas: transportation flows, wildfire evolution, the spatio-temporal evolution of storms, and the spread of infectious diseases.

Booker, Dennis, Frank, Serafini and Torczon (1997) describe a project to design an optimally shaped helicopter blade. While the thrust of their report concerned the development of an optimization algorithm that was used to minimize a function of the computer model outputs, their application is of interest because the engineering specification of the rotor required 31 design variables. Their specific objective function was a measure of the rotor vibration that combined the forces and moments on the rotor, these latter quantities being calculated by the computer code. Each run of the computer code required very little time (10-15 minutes). However, the

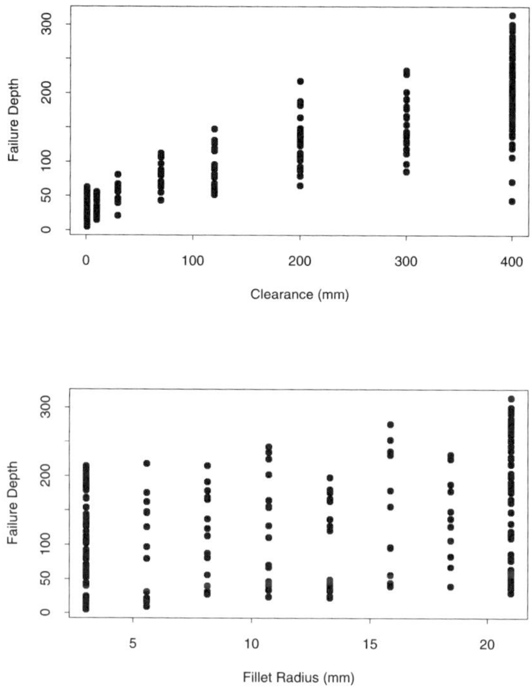

FIGURE 1.6. Top panel—scatterplot of failure depth (millimeters) versus clearance for 234 runs of the computer code described in Subsection 1.2.1; Bottom panel—failure depth versus fillet radius for same data.

computer code provided a much less accurate solution of the mathematical equations that describe the forces and moments on the rotor than did the finite element code of Chang et al. (2001) for their application. This circumstance raises the question of how a fast and slower, gold standard code for the same output can be combined. This issue will be addressed in Section 5.2. Section 7.2 will provide other information about statistical approaches to combining information from multiple sources.

We end this section with a description of how computer codes have played an important role in public policy decision making. Lempert, Schlensinger, Bankes and Andronova (2000) provide an example of such application. The objective is to contrast the effects of several national policies for curbing the effect of greenhouse gases based on an "integrative" model of the future that links the world economy to the population and to the state of the environment. The model they utilized, the so-called *Wonderland model*, quantifies the state of the future over a window of typically 100 years, using several measures, one of which is a "human development index." The model is integrative in that, for example, the pollution at a given time point depends on the user-specified innovation rate for the pollution abatement, the current population, the output per capita, environmental taxes, and other factors. The human development index is a weighted average of a discounted annual improvement of four quantities including, for example, the (net) output per capita. There are roughly 30 input variables. Different public policies can be specified by some of the inputs while the remaining input variables determine the different initial conditions and evolution rates.

1.3 Organization of the Book

The remainder of the book is organized as follows. Chapter 2 outlines the conceptual framework for thinking about the design and analysis of computer experiments. This includes a classification of the types of input variables that can affect the output of a computer code, a summary of research goals when conducting a computer experiment, and an introduction to Gaussian random field models as a description of the output from a computer experiment. Using the Gaussian random model, Chapter 3 introduces methods that can be used for predicting the output of computer codes based on training data, plus methods of assessing the uncertainty in these predictions. The chapter compares the prediction methods and presents our recommendations concerning their use. Chapter 4 introduces several additional topics including the use of predictive distributions and prediction based on multiple outputs. Chapter 5 and Chapter 6 are concerned with experimental design, i.e., the selection of the input sites at which to run code. Chapter 5 begins this discussion by considering space-filling designs, meaning designs that spread observations evenly throughout the input region. Among the designs examined are those based on simple random sampling, those based on stratified random sampling, Latin hypercube designs, orthogonal arrays, distance-based designs, uniform designs, and designs that combine multiple criteria. Sobol´ designs, grid, and lattice designs are briefly mentioned at the end of the chapter. Chapter 6 considers designs based on statistical criteria such as maximum entropy and mean squared error of prediction. This chapter also considers sequential strategies for designing computer experiments when the goal is to optimize the output. Finally, Chapter 7 discusses some issues of validation of computer experiments using physical and other experiments as well as sensitivity analysis.

`PErK` software allows readers to fit most of the models discussed in this book. `PErK`, written in C and using the freely available GSL C software library, can be obtained at either

`http://www.stat.ohio.edu/~comp_exp`

`http://www.springer-ny.com/`

Appendix C describes the syntax used by `PErK` and provides examples of its use to fit a variety of models.

2
Preliminaries

2.1 Introduction

This chapter outlines the basic considerations in thinking about the design and analysis of computer experiments. This section begins by distinguishing *three* types of variables that can affect the output of a computer code $y(\cdot)$, depending on the phenomenon being modeled. Using this categorization, we identify some possible experimental goals.

The first type of variable that we distinguish is a *control variable.* If the output of the computer experiment is some performance measure of a product or process, then the control variables are those variables that can be set by an engineer or scientist to "control" the product or process. Some authors use the terms *engineering variables* or *manufacturing variables* rather than control variables. We use the generic notation $\boldsymbol{x}_c$ to denote control variables. Control variables are present in physical experiments as well as in many computer experiments.

As examples of control variables, we mention the dimensions b and d of the bullet tip prosthesis illustrated in Figure 1.2 (see Subsection 1.2.2). Another example is given by Box and Jones (1992) in the context of a hypothetical physical experiment to formulate ("design") the recipe for a cake. The goal was to determine the amounts of three baking variables to produce the best tasting cake: *flour*, *shortening*, and *egg*; hence, these are control variables. The physical experiment considered two additional variables that also affect the taste of the final product: the *time* at which the cake is baked and the *oven temperature.* Both of the latter variables are specified in the baking recipe on the cake box. However, not all bakers follow the box directions exactly and even if they attempt to follow them precisely, ovens can have true temperatures that differ from their nominal settings and timers can be systematically off or be unheard when they ring.

The variables, baking time and oven temperature, are examples of *environmental variables*, a second type of variable that can be present in both computer and physical experiments. In general, environmental variables affect the output $y(\cdot)$ but depend on the specific user or on the environment at the time the item is used. Environmental variables are sometimes called *noise variables.* We use the notation $\boldsymbol{x}_e$ to denote the vector of environmental variables for a given problem. In practice, we typically regard environmental variables as random with a distribution that is known or unknown. To emphasize situations where we regard the environmental variables as random, we use the notation $\boldsymbol{X}_e$. The hip prosthesis example of Chang, Williams, Notz, Santner and Bartel (1999) illustrates a computer experiment with environmental variables (see Subsection 1.2.2); both of their outputs depended on the *magnitude* and *direction* of the force exerted on the head of the prosthesis. These two variables were patient specific and depended on body mass and activity. They were treated as having a given distribution that was characteristic of a given population.

In addition to control and environmental variables, there is a third category of input variable that sometimes occurs. This third type of input variable describes the uncertainty in the mathematical modeling that relates other inputs to output(s). As an example, O'Hagan, Kennedy and Oakley (1999) consider a model for metabolic processing of U^{235} that involves various rate constants for elementary conversion processes that must be known in order to specify the overall metabolic process. In some cases, such elementary rate constants may have values that are unknown or possibly there is a known (subjective) distribution that describes their values. We call these variables *model* variables and denote them by $\boldsymbol{x}_m$. In a classical statistical setting we would call model variables "model parameters" because we use the results of a physical experiment, the ultimate reality, to estimate their values. Some authors call model variables "tuning parameters."

The following section describes several fundamental goals for computer experiments depending on which types of variables are present and the number of responses that the code produces. For example, if the code produces a single real-valued response that depends on control and environmental variables, then we use the notation $y(\boldsymbol{x}_c, \boldsymbol{X}_e)$ to emphasize that the propagation of uncertainty in the environmental variables $\boldsymbol{X}_e$ must be accounted for. In some cases there may be multiple computer codes that produce related responses $y_1(\cdot), \ldots, y_m(\cdot)$ which either represent competing responses or correspond to "better" and "worse" approximations to the response. For example, if there are multiple finite element analysis codes based on greater or fewer node/edge combinations to represent the *same* phenomenon, then one might hope to combine the responses to improve prediction. Another alternative is that $y_1(\cdot)$ represents the primary object of interest while $y_2(\cdot), \ldots, y_m(\cdot)$ represent "related information"; for example, this would be the case if the code produced a response *and* vector of first partial derivatives. A third possibility is when the $y_i(\cdot)$ represent competing objectives; in this case, the goal might be to optimize one response subject to minimum performance standards on the remaining ones.

Following the description of experimental goals, we summarize the basic issues in modeling computer output. Then we will be prepared to begin Chapter 3 on the first of the two basic issues considered in this book, that of predicting $y(\cdot)$ at (a new) input $\boldsymbol{x}_0$ based on training data $(\boldsymbol{x}_1, y(\boldsymbol{x}_1)), \ldots, (\boldsymbol{x}_n, y(\boldsymbol{x}_n))$. Chapter 5 will address the second issue, the design problem of choosing the input sites at which the computer model should be run.

2.2 Defining the Experimental Goal

2.2.1 *Introduction*

In this section, we initially consider the case of a single real-valued output $y(\cdot)$ that is to be evaluated at input training sites $\boldsymbol{x}_1, \ldots, \boldsymbol{x}_n$. We let $\widehat{y}(\boldsymbol{x})$ denote a generic predictor of $y(\boldsymbol{x})$ and consider goals for two types of inputs. In the first type of input, referred to as a *homogeneous-input*, all components of $\boldsymbol{x}$ are either control variables *or* environmental variables *or* model variables. In the second type of input, referred to as a *mixed-input*, $\boldsymbol{x}$ contains at least two of the three different types of input variables: control, environmental, and model. Finally, in Subsection 2.2.4, we outline some typical goals when there are several outputs. In all cases there can be both "local" and "global" goals that may be of interest.

2.2.2 *Research Goals for Homogeneous-Input Codes*

First, suppose that $\boldsymbol{x}$ consists exclusively of control variables, i.e., $\boldsymbol{x} = \boldsymbol{x}_c$. In this case one important objective is to predict $y(\boldsymbol{x})$ "well" for all $\boldsymbol{x}$ in some domain $\mathcal{X}$. There have been several criteria used to measure the quality of the prediction in an "overall" sense. One appealing intuitive basis for judging the predictor $\widehat{y}(\boldsymbol{x})$ is its *integrated squared error*

$$\int_{\mathcal{X}} \left[\widehat{y}(\boldsymbol{x}) - y(\boldsymbol{x})\right]^2 w(\boldsymbol{x})\, d\boldsymbol{x}, \tag{2.2.1}$$

where $w(\boldsymbol{x})$ is a nonnegative weight function that quantifies the importance of each value in $\mathcal{X}$. For example, $w(\boldsymbol{x}) = 1$ weights all parts of $\mathcal{X}$ equally while $w(\boldsymbol{x}) = I_{\mathcal{A}}(\boldsymbol{x})$, the indicator function of the set $\mathcal{A} \subset \mathcal{X}$, ignores the complement of $\mathcal{A}$ and weights all points in $\mathcal{A}$ equally.

Unfortunately, (2.2.1) cannot be calculated because $y(\boldsymbol{x})$ is unknown. However, later in Chapter 6 we will replace $[\widehat{y}(\boldsymbol{x}) - y(\boldsymbol{x})]^2$ by a posterior mean squared value computed under a certain "prior" model for $y(\boldsymbol{x})$ and obtain a quantity that can be computed (see Section 6.2 for methods of designing computer experiments in such settings).

The problem of predicting $y(\cdot)$ well over a region can be thought of as a global objective. In contrast, more local goals focus on finding "interesting" parts of the input space $\mathcal{X}$. An example of such a goal is to identify (any) $\boldsymbol{x}$, where $y(\boldsymbol{x})$ equals some target value. Suppose

$$\mathcal{L}(t_0) = \{\boldsymbol{x} \in \mathcal{X} \mid y(\boldsymbol{x}) = t_0\}$$

denotes the "level set" of input values where $y(\cdot)$ attains a target value t_0. Then we wish to determine any input $\boldsymbol{x}$ where $y(\cdot)$ attains the target level, i.e., any $\boldsymbol{x} \in \mathcal{L}(t_0)$. Another example of a local goal is to find extreme values of $y(\cdot)$. Suppose

$$\mathcal{M} = \{\boldsymbol{x} \in \mathcal{X} \mid y(\boldsymbol{x}) \geq y(\boldsymbol{x}^\star) \text{ for all } \boldsymbol{x}^\star \in \mathcal{X}\} \equiv \arg\max y(\cdot)$$

is the set of all arguments that attain the global maximum of $y(\boldsymbol{x})$. Then an analog of the level set problem is to find a set of inputs that attain the overall maximum, i.e., to determine any $\boldsymbol{x} \in \mathcal{M}$. The problem of finding global optima of computer code output has been the subject of much investigation (Mockus, Tiešis and Žilinskas (1978), Bernardo, Buck, Liu, Nazaret, Sacks and Welch (1992), Mockus, Eddy, Mockus, Mockus and Reklaitis (1997), Jones, Schonlau and Welch (1998), Schonlau, Welch and Jones (1998)).

There is a large literature on homogeneous-input problems when $\boldsymbol{x}$ depends only on environmental variables. Perhaps the most frequently occurring application is when the environmental variables are random inputs with a known distribution and the goal is determine how the variability in the inputs is transmitted through the computer code. In this case we write $\boldsymbol{x} = \boldsymbol{X}_e$ using upper case notation to emphasize that the inputs are to be treated as random variables and the goal is that of finding the distribution of $y(\boldsymbol{X}_e)$. This problem is sometimes called uncertainty analysis (Crick, Hofer, Jones and Haywood (1988), Dandekar and Kirkendall (1993), Helton (1993), O'Hagan and Haylock (1997), and O'Hagan et al. (1999) are examples of such papers). Also in this spirit, McKay, Beckman and Conover (1979) introduced the class of Latin hypercube designs for choosing the training sites $\boldsymbol{X}_e$ at which to evaluate the code when the problem is to predict the *mean* of the $y(\boldsymbol{X}_e)$ distribution, $E\{y(\boldsymbol{X}_e)\}$. The theoretical study of Latin hypercube designs has established a host of asymptotic and empirical properties of estimators based on them (Stein (1987), Owen (1992a), Owen (1994), Loh (1996), Pebesma and Heuvelink (1999)) and enhancements of such designs (Handcock (1991), Tang (1993), Tang (1994), Ye (1998), Butler (2001)).

The third possibility for homogeneous-input is when $y(\cdot)$ depends only on *model* variables, $\boldsymbol{x} = \boldsymbol{x}_m$. Typically in such a case, the computer code

then use these models to derive (empirical) best linear unbiased predictors of $y(\cdot)$ at new sites $\boldsymbol{x}_0$ based on all the responses. See Section 4.2 for a discussion of modeling multiple responses.

Now consider the scenario where $\boldsymbol{x} = \boldsymbol{x}_c$, $y_1(\cdot)$ is the response of primary interest, and $y_2(\cdot), \ldots, y_m(\cdot)$ are competing objectives. Then we can define a feasible region of $\boldsymbol{x}_c$ values by requiring minimal performance standards for $y_2(\boldsymbol{x}_c), \ldots, y_m(\boldsymbol{x}_c)$. Formally, an analog of the problem of minimizing $y(\cdot)$ is

$$\begin{array}{ccc} \text{minimize} & y_1(\boldsymbol{x}_c) & \\ & \text{subject to} & \\ y_2(\boldsymbol{x}_c) & \geq & M_2 \\ & \vdots & \\ y_m(\boldsymbol{x}_c) & \geq & M_m. \end{array}$$

Here M_i is the lower bound on the performance of $y_i(\cdot)$ that is acceptable. If in addition to control variables, $\boldsymbol{x}$ also contains environmental variables, then we can replace each $y_i(\boldsymbol{x}_c)$ above with $\mu_i(\boldsymbol{x}_c) = E\{y_i(\boldsymbol{x}_c, \boldsymbol{X}_e)\}$. In cases where $\boldsymbol{x} = \boldsymbol{x}_e$, a typical objective is to find the joint distribution of $(y_1(\boldsymbol{X}_e), \ldots, y_m(\boldsymbol{X}_e))$ or, even simpler, that of predicting the mean vector $(E\{y_1(\boldsymbol{X}_e)\}, \ldots, E\{y_m(\boldsymbol{X}_e)\})$.

Lastly, if the $y_i(\cdot)$ represent the outputs of *different* codes of varying accuracy for the same response, then a typical goal is to combine information from the various outputs to better predict the true response. Specification of this goal depends on identifying the "true" response; we postpone a discussion of this idea until we discuss modeling multiple response output in Section 4.2.

2.3 Modeling Output from Computer Experiments

2.3.1 *Introduction*

This book uses Bayesian methodology to design and analyze computer experiments. Prior information describing the functional relationship of the input $\boldsymbol{x}$ to the (unknown) output $y(\boldsymbol{x})$ is combined with the information in the training data to predict $y(\cdot)$ at new sites and to accomplish the other goals described in Sections 2.2 and 6.3.

Best linear unbiased prediction, a frequentist methodology, has also been used for prediction of real-valued quantities associated with $y(\cdot)$ and for the calculation of their standard errors (see Subsection 3.2.3). The conceptual problem with this approach is that the source of randomness that is measured by the standard error, for example, is not easily understood and when specified is often not of interest to the user. For example, one source of randomness that leads to interpretable standard errors is the randomness in the predictor that results from use of a stochastic mechanism to choose the locations of the input data (the "design" of the computer experiment). In this case, the standard error of a predictor of $y(\boldsymbol{x}_0)$ is the variation in the predictor due to the randomly selected training data.

We prefer the Bayesian approach to analyze the data from computer experiments and regard the use of a prior distribution for $y(\cdot)$ as clearer in its intent than the frequentist viewpoint, though not simpler to implement. Computer experiments represent a highly nonparametric setting. Eliciting a prior for the output of a black box code is much more difficult than, say, eliciting the prior for the output of a regression. However, this approach is philosophically more satisfying, for example, in its interpretation of the standard errors that will be specified in Section 4.1—they refer to model uncertainty (given the training data). However, the reader should recognize that reasonable users may disagree about the prior information concerning the input-output function that any particular Bayesian predictor makes (and hence the associated standard error of prediction). Oakley (2002) and Reese, Wilson, Hamada, Martz and Ryan (2000) give advice and case-studies about the formation of prior distributions.

In sum, our attitude toward using the Bayesian approach to problems of the design and analysis of computer experiments is not dogmatic. We *do* attempt to control the characteristics of the functions produced by our priors, but *do not* rigidly believe them. Instead, our goal is to choose flexible priors that are capable of producing many shapes for $y(\cdot)$ and then let the Bayesian machinery allow the data to direct the details of the prediction process.

Subsections 2.3.2-2.3.4 will introduce Gaussian random functions and provide the reader with an appreciation for the flexibility of this class of priors. Subsection 2.3.5 will discuss hierarchical priors based on Gaussian random functions as a method of further enhancing this flexibility.

The final general point we wish to make in this introduction is that computer experiments are not alone in their use of Bayesian prediction methodology to analyze high-dimensional, highly correlated data. Many other scientific fields produce such data, albeit with measurement error. The statistical analyses used in geostatistics (Matheron (1963), Journel and Huijbregts (1979)), environmental statistics and disease mapping (Ripley (1981), Cressie (1993)), global optimization (Mockus et al. (1997)), and statistical learning (Hastie, Tibshirani and Friedman (2001)) are based on the Bayesian philosophy. Hence many of the methodologies discussed in their literatures are also relevant here.

In the following we regard $y(\cdot)$ to be a real-valued function with domain $\mathcal{X}$ where $\mathcal{X}$ is a subset of d-dimensional Euclidean space having positive d-dimensional volume. We adopt the notation $Y(\cdot)$ to distinguish the random function from its realizations $y(\cdot)$ which are functions. Some authors use the terms "stochastic process" or simply "process" rather than random function and we occasionally also use these terms, although "random function" is the most natural terminology when discussing computer experiments.

Conceptually, a random function should be thought of as a mapping from elements of a sample space of outcomes, say Ω, to a given set of functions, just as random variables are mappings from a set Ω of elementary outcomes to the real numbers. It will occasionally add clarity to our discussion to make this explicit by writing $y(\boldsymbol{x}) = Y(\boldsymbol{x}, \omega)$ to be a *particular* function from $\mathcal{X}$ to $\mathbb{R}^1$, where $\omega \in \Omega$ is a specific element in the sample space. Sometimes we refer to $y(\cdot, \omega)$ as a *draw* from the random function $Y(\cdot)$ or as a *sample path* (in $\mathcal{X}$) of the random function. The introduction of the underlying sample space Ω helps clarify ideas when discussing the smoothness properties of functions drawn from $Y(\cdot)$. In particular, we desire sufficient flexibility in our stochastic model so that, ideally, there is an ω for which $y(\boldsymbol{x}) = Y(\boldsymbol{x}, \omega)$ represents the response to our computer experiment.

We will also consider computer experiments that produce multiple outputs; in such situations we let $\boldsymbol{y}(\boldsymbol{x}) = (y_1(\boldsymbol{x}), \ldots, y_m(\boldsymbol{x}))^\top$ denote the vector of outputs. A typical application that produces multiple outputs is computer code that determines not only $y(\boldsymbol{x})$ but also each of the partial derivatives of $y(\boldsymbol{x})$. Then $\boldsymbol{y}(\boldsymbol{x}) = (y(\boldsymbol{x}), \partial y(\boldsymbol{x})/\partial x_1, \ldots, \partial y(\boldsymbol{x})/\partial x_d)$. In the general multiple output case, we view the random mechanism as associating a vector valued function, $\boldsymbol{y}(\boldsymbol{x}) = \boldsymbol{Y}(\boldsymbol{x}, \omega)$, with each elementary outcome $\omega \in \Omega$. Codes that produce multiple outcomes were introduced in Section 2.2; their modeling will be considered in Subsection 5.2.1; applications of such models will be provided in Subsections 4.2.3 and 6.3.6.

We begin this overview of stochastic models for generating functions $y(\cdot)$ with the following simple example.

Example 2.1 Suppose that we generate $y(x)$ on $[-1, +1]$ by the mechanism

$$Y(x) = b_0 + b_1 x + b_2 x^2, \tag{2.3.1}$$

where b_0, b_1, and b_2 are independent with $b_i \sim N(0, \sigma_i^2)$ for $i = 1, 2, 3$. Functions drawn from $Y(x)$ are simple to visualize. Every realization $y(\cdot)$ is a quadratic equation ($P\{b_2 = 0\} = 0$) that is symmetric about an axis other than the y-axis (symmetry about the y-axis occurs if and only if $b_1 = 0$ and $P\{b_1 = 0\} = 0$). The quadratic is convex with probability 1/2 and it is concave with probability 1/2 (because $P\{b_2 > 0\} = 1/2 = P\{b_2 < 0\}$). Figure 2.1 illustrates ten outcomes from this random function when $\sigma_0^2 = \sigma_1^2 = \sigma_2^2 = 1.0$.

For any $x \in [-1, +1]$ the draws from (2.3.1) have mean zero, i.e.,

$$\begin{aligned} E\{Y(x)\} &= E\{b_0 + b_1 x + b_2 x^2\} \\ &= E\{b_0\} + E\{b_1\} \times x + E\{b_2\} \times x^2 \\ &= 0 + 0 \times x + 0 \times x^2 = 0. \end{aligned} \tag{2.3.2}$$

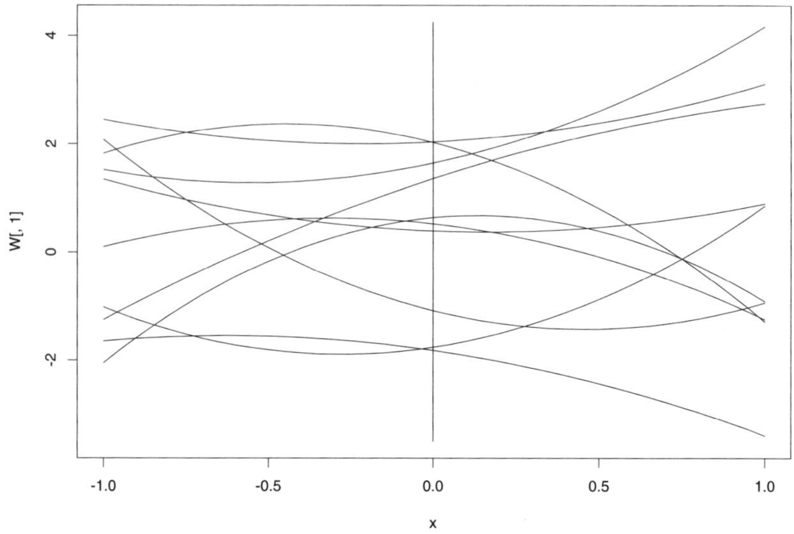

FIGURE 2.1. Ten draws from the random function $Y(x) = b_0 + b_1x + b_2x^2$ on $[-1,+1]$, where b_0, b_1, and b_2 are independent and identically $N(0,1.0)$ distributed.

Equation (2.3.2) says that for any x, the mean of $Y(x)$ is *zero* over many drawings of the coefficients (b_0,b_1,b_2); this is true because each regression coefficient is independent and centered at the origin so that each regression term is positive and negative with probability 1/2 and thus their sum, $Y(x)$, is also positive and negative with probability 1/2.

For any $x \in [-1,+1]$ the pointwise variance of $Y(x)$ is

$$\begin{aligned} \mathrm{Var}\{Y(x)\} &= E\left\{\left(b_0+b_1x+b_2x^2\right)\left(b_0+b_1x+b_2x^2\right)\right\} \\ &= \sigma_0^2+\sigma_1^2x^2+\sigma_2^2x^4 \geq 0. \end{aligned}$$

The values of $Y(x_1)$ and $Y(x_2)$ at x_1, $x_2 \in [-1,+1]$ are related, as can be seen from

$$\begin{aligned} \mathrm{Cov}\{Y(x_1),Y(x_2)\} &= E\left\{\left(b_0+b_1x_1+b_2x_1^2\right)\left(b_0+b_1x_2+b_2x_2^2\right)\right\} \\ &= \sigma_0^2+\sigma_1^2x_1x_2+\sigma_2^2x_1^2x_2^2. \end{aligned} \tag{2.3.3}$$

This covariance can be positive or negative. The sign of the covariance of $Y(x_1)$ and $Y(x_2)$ can intuitively be explained as follows. The covariance formula (2.3.3) is clearly positive for any x_1 and x_2 when both are positive or both are negative. Intuitively this is true because over many drawings of (b_0,b_1,b_2), x_1 and x_2 both tend to be on the same side of the axis of symmetry of the quadratic and thus $Y(x_1)$ and $Y(x_2)$ increase or decrease together. The covariance formula *can* be negative if x_1 and x_2 are on the *opposite* sides of the origin *and* σ_1^2 dominates σ_0^2 and σ_2^2 (algebraically, the middle term in (2.3.3) is negative and can exceed the sum of the other two terms). Intuitively, one circumstance where this occurs is if σ_0^2 is small (meaning the curves tend to go "near" $(0, 0)$), *and* σ_2^2 is small (the curves tend to be linear near the origin), *and* σ_1^2 is large; in this case, the draws fluctuate between those with large positive slopes and those with large negative slopes, implying that $Y(x_1)$ and $Y(x_2)$ tend to have the opposite sign over the draws.

Because linear combinations of a fixed set of independent normal random variables have the normal distribution, the simple model (2.3.1) for $Y(\cdot)$ satisfies: for each $L > 1$ and any choice of $x_1, \ldots, x_L \in \mathcal{X}$, the vector $(Y(x_1), \ldots, Y(x_L))$ is multivariate normally distributed. (See Appendix B for a review of the multivariate normal distribution.) The $y(\cdot)$ realizations have several limitations from the viewpoint of computer experiments. First, the model can *only* produce quadratic draws. Second, the multivariate normal distribution of $(Y(x_1), \ldots, Y(x_L))$ is *degenerate* when $L \geq 4$. In the development below we wish to derive more flexible random functions that retain the computational advantage that $(Y(x_1), \ldots, Y(x_L))$ has the multivariate normal distribution. ■

There are many sources that provide detailed theoretical discussions of random functions, particularly the Gaussian random functions introduced in Subsections 2.3.2–2.3.4 (Cramér and Leadbetter (1967), Adler (1981), Adler (1990), and Abrahamsen (1997), for example). It is not our purpose to present a complete account of the theory. Rather, we desire to give an overview of these models, to describe the relationship between the "correlation function" of stationary Gaussian random functions and the smoothness properties of its realizations $y(\boldsymbol{x})$, and to develop intuition about this relationship through a series of examples.

2.3.2 Gaussian Random Function Models

In the computer experiments literature, the most popular models for generating function draws are *Gaussian random functions*, also called the Gaussian stochastic processes. Hence we emphasize these models in this section although, as we will note, some of the concepts that we introduce apply to more general random functions.

Definition Suppose that $\mathcal{X}$ is a fixed subset of $\mathbb{R}^d$ having positive d-dimensional volume. We say that $Y(\boldsymbol{x})$, for $\boldsymbol{x} \in \mathcal{X}$, is a *Gaussian random function* (GRF) provided that for any $L \geq 1$ and any choice of $\boldsymbol{x}_1, \ldots, \boldsymbol{x}_L$ in $\mathcal{X}$, the vector $(Y(\boldsymbol{x}_1), \ldots, Y(\boldsymbol{x}_L))$ has a multivariate normal distribution.

Gaussian random functions are determined by their *mean* function, $\mu(\boldsymbol{x}) \equiv E\{Y(\boldsymbol{x})\}$, for $\boldsymbol{x} \in \mathcal{X}$, and by their *covariance* function

$$C^{\star}(\boldsymbol{x}_1, \boldsymbol{x}_2) \equiv \text{Cov}\{Y(\boldsymbol{x}_1), Y(\boldsymbol{x}_2)\},$$

for $\boldsymbol{x}_1, \boldsymbol{x}_2 \in \mathcal{X}$. Some authors call $C^{\star}(\cdot, \cdot)$ the "autocovariance" function to be consistent with the language used in time series analysis.

The $Y(\boldsymbol{x})$ model in Example 2.1 is a GRF. The GRFs that are used in practice are *nonsingular*, which means that for any choice of inputs, the covariance matrix of the associated multivariate normal distribution is nonsingular. Such nonsingular multivariate normal distributions have the advantage that it is easy to compute the conditional distribution of one (or several) of the $Y(\boldsymbol{x}_i)$ variables given the remaining $Y(\boldsymbol{x}_j)$. The prediction methodology used in Section 3.3 requires that these conditional means and conditional variances be known and the predictive distributions of Section 4.1 require that the entire conditional distribution be known. In addition, draws from the most widely used GRFs allow a greater spectrum of shapes than the quadratic equations generated in (2.1). They also permit the modeler to control the smoothness properties of the $y(\boldsymbol{x})$ draws; in most of the scientific applications mentioned above, there is *some* information about the smoothness of $y(\cdot)$, although perhaps only that it is a continuous function of the inputs.

There are two technical concepts that we address briefly before introducing specific GRF models. We wish to make the reader aware of the *practical* difficulties that these two concepts address. The first concept has to do with the fact that our random function models are defined by their *finite-dimensional* distributions while, in the following, we are interested in properties that depend on limiting operations such as assuring that functions drawn from the process have specified smoothness (continuity and differentiability) properties. The continuity and differentiability of $y(\boldsymbol{x})$ as a function of $\boldsymbol{x}$ are *sample path properties*, i.e., they regard $y(\boldsymbol{x}) = Y(\boldsymbol{x}, \omega)$ as a function of $\boldsymbol{x}$ for fixed ω. Thus throughout, we require that our random function models be *separable*, which is a property introduced by Doob (1953) that ensures that the finite-dimensional distributions determine the sample path properties of function draws. Adler (1981) (pages 14-15) states the formal definition of separability and discusses its intuition. For our purposes it suffices to know that given any random function $Y(\cdot)$ on $\mathcal{X}$, there is an equivalent separable random function $Y^s(\cdot)$ on $\mathcal{X}$. The random functions $Y(\cdot)$ and $Y^s(\cdot)$ are *equivalent* provided

$$P\{Y(\boldsymbol{x}) = Y^s(\boldsymbol{x})\} = 1 \quad \text{for all} \quad \boldsymbol{x} \in \mathcal{X}.$$

We assume throughout (and the proofs of almost sure sample path properties require) that our GRF models have been chosen to be separable.

Our second technical concept concerns a statistical issue. Classical statistical methods make inferences about a population based on a random sample of data from that population. Indeed, the statistical procedure is chosen to have certain sampling characteristics (meaning properties based on repeated sampling from the population). In computer experiments (as well as in most other applications of spatial statistics), we observe $y(\boldsymbol{x}_1), \ldots, y(\boldsymbol{x}_n)$, where $\boldsymbol{x}_1, \ldots, \boldsymbol{x}_n$ are training data input sites. However, these data are values of a *single* function drawn from a population of functions according to $Y(\cdot)$, i.e., $(y(\boldsymbol{x}_1), \ldots, y(\boldsymbol{x}_n)) = (Y(\boldsymbol{x}_1, \omega), \ldots, Y(\boldsymbol{x}_n, \omega))$. Thus spatial data gives *partial* information about a single function $y(\boldsymbol{x}) = Y(\boldsymbol{x}, \omega)$ rather than a *random sample of functions* drawn according to $Y(\cdot)$. To predict the value of $y(\boldsymbol{x}_{new})$, where $\boldsymbol{x}_{new}$ is a new input site, the process must exhibit some regularity over $\mathcal{X}$. In general, it need not be the case that one can make inference about population quantities which are Ω averages such as the $y(\boldsymbol{x}_{new})$ predictor above, based on a spatial average for a single $\omega \in \Omega$. Process *ergodicity* is the standard property that permits valid statistical inference about that process based on a single draw (for a discussion of this property from a statistical viewpoint, see Cressie (1993), pages 52-58, and the additional references listed there). The technical details of this concept are beyond the scope of this book; we note only that this issue motivates users to restrict attention to GRFs that are (strongly) *stationary* (or homogeneous).

Definition The random function $Y(\cdot)$ is *strongly stationary* provided that for any $\boldsymbol{h} \in \mathbb{R}^d$, any $L \geq 1$, any $\boldsymbol{x}_1, \ldots, \boldsymbol{x}_L$ in $\mathcal{X}$ with $\boldsymbol{x}_1 + \boldsymbol{h}, \ldots, \boldsymbol{x}_L + \boldsymbol{h} \in \mathcal{X}$, it must be the case that $(Y(\boldsymbol{x}_1), \ldots, Y(\boldsymbol{x}_L))$ and $(Y(\boldsymbol{x}_1 + \boldsymbol{h}), \ldots, Y(\boldsymbol{x}_L + \boldsymbol{h}))$ have the *same* distribution.

Notice that this definition is general. When applied to GRFs $Y(\cdot)$, stationarity is equivalent to requiring that $(Y(\boldsymbol{x}_1), \ldots, Y(\boldsymbol{x}_L))$ and $(Y(\boldsymbol{x}_1 + \boldsymbol{h}), \ldots, Y(\boldsymbol{x}_L + \boldsymbol{h}))$ always have the same mean vector and same covariance matrix. In particular, GRFs must have the *same* marginal distribution for all $\boldsymbol{x}$ (taking $L = 1$); their mean and their variance must both be constant. Furthermore, it is not difficult to show that the covariance of a stationary GRF must satisfy

$$\text{Cov}\left\{Y(\boldsymbol{x}_1), Y(\boldsymbol{x}_2)\right\} = C\left(\boldsymbol{x}_1 - \boldsymbol{x}_2\right) \tag{2.3.4}$$

for some function $C(\cdot)$, called the *covariance function* of the process. The equation (2.3.4) means that all pairs of locations $\boldsymbol{x}_1$ and $\boldsymbol{x}_2$ having common orientation *and* common inter-point distance will have the same covariance. For example, the pairs of points at the tails and tips of the three arrows in Figure 2.2 all have the same covariance structure (as well as infinitely many other pairs on the two parallel lines depicted in the figure). The (constant) variance of a stationary process can be expressed in terms of its covariance function as $\text{Var}\{Y(\boldsymbol{x})\} = \text{Cov}\{Y(\boldsymbol{x}), Y(\boldsymbol{x})\} = C(\boldsymbol{0})$.

Technically, the stationarity of a GRF $Y(\boldsymbol{x})$ does not guarantee that $Y(\boldsymbol{x})$ is ergodic but this will be case if $C(\boldsymbol{h}) \to 0$ as $\boldsymbol{h} \to \infty$ and *hence, inference is valid based on data collected from a single sample path* (Adler

(1981), page 145). The correlation function examples below satisfy this condition.

An even stronger requirement is that the GRF be invariant under rotations, a property called *isotropy*. A stationary GRF $Y(\cdot)$ can be shown to be isotropic provided

$$\mathrm{Cov}\left\{Y(\boldsymbol{x}_1), Y(\boldsymbol{x}_2)\right\} = C\left(\|\boldsymbol{x}_1 - \boldsymbol{x}_2\|_2\right), \qquad (2.3.5)$$

where $\|\boldsymbol{h}\|_2 = \sqrt{\sum_i h_i^2}$ is Euclidean distance. For isotropic models, every pair of points $\boldsymbol{x}_1$ and $\boldsymbol{x}_2$ having common inter-point distance must have the same covariance (and correlation) regardless of their orientation (see the right panel of Figure 2.2). For example, for any isotropic GRF, the origin has the same correlation with every point on the unit circle. Isotropic models are usually not useful when component inputs are measured on different scales.

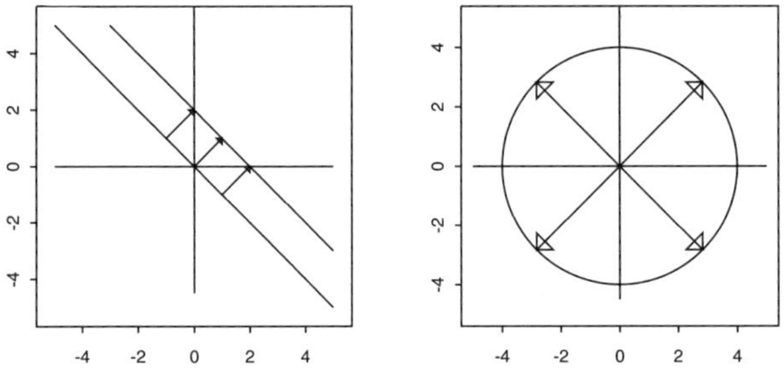

FIGURE 2.2. In the left-hand panel, the tip and tail of each arrow have the same correlation for stationary random functions. In the right-hand panel, all points on the circle have the same correlation for isotropic random functions.

We will occasionally consider random function models that are nonparametric in that they make only moment requirements on $Y(\cdot)$. The most important such model is that of *second-order stationary*. A random function $Y(\cdot)$ having *constant mean* and *constant variance* is second-order stationary provided its covariance function satisfies (2.3.4).

Despite the arguments given above, stationarity is a substantial restriction and we often require more flexibility in modeling $y(\boldsymbol{x})$. Several approaches have been used in the literature to enhance random function modeling while retaining (some) of the theoretical simplifications that stationarity provides. The most frequently used of these techniques, the one we employ here, is to permit the mean of the stochastic process generating $y(\boldsymbol{x})$ to depend on $\boldsymbol{x}$ in a standard regression manner while assuming the residual variation follows a stationary GRF. The corresponding random function has the form

$$Y(\boldsymbol{x}) = \sum_{j=1}^{p} f_j(\boldsymbol{x})\beta_j + Z(\boldsymbol{x}) = \boldsymbol{f}^\top(\boldsymbol{x})\boldsymbol{\beta} + Z(\boldsymbol{x}), \qquad (2.3.6)$$

where $f_1(\cdot), \ldots, f_p(\cdot)$ are *known* regression functions, $\boldsymbol{\beta} = (\beta_1, \ldots, \beta_p)^\top$ is a vector of *unknown* regression coefficients, and $Z(\cdot)$ is a *zero mean* stationary GRF over $\mathcal{X}$. These $Y(\cdot)$ models are, of course, nonstationary.

Most other methods for enhancing $Y(\boldsymbol{x})$ model flexibility have been motivated by environmental applications which often require nonstationary models. While extremely successful in these applications, the nonstationary models introduced in the course of such data analyses have typically been used only in low-dimensional $\boldsymbol{x}$ input settings (two- or three-dimensional space, or three- or four-dimensional space-time applications). Their ability to handle higher dimensional $\boldsymbol{x}$ input cases is untested, although they may well be of use in the analysis of computer experiments. We mention two modeling strategies that have been suggested in the literature.

One method is to generate nonstationary $Y(\boldsymbol{x})$ models from stationary ones by convolving a stationary process with a kernel; Higdon, Swall and Kern (1999) integrate white noise, the spatial analog of a random sample of normal observations, against a Gaussian kernel to produce such a model. In the same spirit, Hass (1995) constructs $Y(\boldsymbol{x})$ models as a moving window over a stationary process. Another approach, introduced by Sampson and Guttorp (1992), is based on deforming the input $\boldsymbol{x}$ of a stationary process to model $Y(\boldsymbol{x})$ (see also Guttorp and Sampson (1994) and Guttorp, Meiring and Sampson (1994)).

2.3.3 The Correlation Function of a Gaussian Random Function Model

To be consistent with the notation introduced in (2.3.6), hereafter we denote the stationary GRF of interest by $Z(\cdot)$. We reiterate that $Z(\cdot)$ has zero mean (by including any overall constant mean value among the regression terms in (2.3.6)). Thus $Z(\cdot)$ is completely determined by its covariance function $C(\cdot)$. In some applications, it is more convenient to separately model the process variance $\sigma_Z^2 = C(\boldsymbol{0})$ and the process correlation function. The *correlation function* of a stationary process $Z(\boldsymbol{x})$ that has finite $\sigma_Z^2 > 0$ and covariance function $C(\cdot)$ is defined to be

$$R(\boldsymbol{h}) = C(\boldsymbol{h})/\sigma_Z^2 \quad \text{for} \quad \boldsymbol{h} \in \mathbb{R}^d.$$

The name "correlation function" comes from

$$\begin{aligned} \text{Cor}\{Z(\boldsymbol{x}_1), Z(\boldsymbol{x}_2)\} &= \frac{\text{Cov}\{Z(\boldsymbol{x}_1), Z(\boldsymbol{x}_2)\}}{\sqrt{\text{Var}\{Z(\boldsymbol{x}_1)\} \times \text{Var}\{Z(\boldsymbol{x}_2)\}}} \\ &= \frac{C(\boldsymbol{x}_1 - \boldsymbol{x}_2)}{\sigma_Z^2} = R(\boldsymbol{x}_1 - \boldsymbol{x}_2). \end{aligned}$$

What properties must valid covariance and correlation functions possess? Assuming that $Z(\boldsymbol{x})$ is nondegenerate, then $C(\boldsymbol{0})$ $(= \sigma_Z^2) > 0$ while $R(\boldsymbol{0}) = 1$. Because $\text{Cov}\{Y(\boldsymbol{x}+\boldsymbol{h}), Y(\boldsymbol{x})\} = \text{Cov}\{Y(\boldsymbol{x}), Y(\boldsymbol{x}+\boldsymbol{h})\}$, the covariance and correlation functions of stationary GRFs must be *symmetric about the origin*, i.e.,

$$C(\boldsymbol{h}) = C(-\boldsymbol{h}) \quad \text{and} \quad R(\boldsymbol{h}) = R(-\boldsymbol{h}).$$

Both $C(\cdot)$ and $R(\cdot)$ must be *positive semidefinite* functions; stated in terms of $C(\cdot)$, this means that for any $L \geq 1$, and any real numbers $w_1, \ldots, w_L$, and any inputs $\boldsymbol{x}_1, \ldots, \boldsymbol{x}_L$ in $\mathcal{X}$,

$$\sum_{i=1}^{L} \sum_{j=1}^{L} w_i w_j C(\boldsymbol{x}_i - \boldsymbol{x}_j) \geq 0. \tag{2.3.7}$$

The sum (2.3.7) must be nonnegative because the left-hand side is the variance of $\sum_{i=1}^{L} w_i Y(\boldsymbol{x}_i)$. The covariance function $C(\cdot)$ is *positive definite* provided >0 holds in (2.3.7) for every $(w_1, \ldots, w_L) \neq \boldsymbol{0}$ (any $L \geq 1$ and any $\boldsymbol{x}_1, \ldots, \boldsymbol{x}_L$ in $\mathcal{X}$).

While every covariance function must satisfy the symmetry and positive semidefinite properties above, these properties do not offer a convenient method for generating valid covariance functions. Rather, what is of greater importance is a characterization of the class of covariance functions because this would allow us to generate valid covariance functions. While a general study of how to determine the form of valid stationary covariance functions is beyond the scope of this book, one answer to this question is relatively simple to state, and we do so next.

As a prelude to identifying this class of covariance functions (and as an introduction to the topic of smoothness which is taken up again in Subsection 2.3.4), we introduce the concept of mean square (MS) continuity. Mean square properties describe the average performance of the sample paths. For purposes of stating the definitions of MS properties, there is nothing to be gained by restricting attention to GRFs and so we consider general random functions $Y(\cdot)$.

Definition Suppose $Y(\cdot)$ is a stationary process on $\mathcal{X}$ that has finite second moments. We say that $Y(\cdot)$ is *MS continuous* at the point $\boldsymbol{x}_0 \in \mathcal{X}$ provided

$$\lim_{\boldsymbol{x} \to \boldsymbol{x}_0} E\left\{(Y(\boldsymbol{x}) - Y(\boldsymbol{x}_0))^2\right\} = 0.$$

The process is *MS continuous on* $\mathcal{X}$ provided it is MS continuous at every $\boldsymbol{x}_0 \in \mathcal{X}$.

Suppose $C_Y(\cdot)$ is the covariance function of the stationary process $Y(\cdot)$, then

$$E\left\{(Y(\boldsymbol{x}) - Y(\boldsymbol{x}_0))^2\right\} = 2\left(C_Y(\boldsymbol{0}) - C_Y(\boldsymbol{x} - \boldsymbol{x}_0)\right). \tag{2.3.8}$$

The right-hand formula shows that $Y(\cdot)$ is MS continuous at $\boldsymbol{x}_0$ provided $C_Y(\cdot)$ is continuous at the origin—in fact, $Y(\cdot)$ is MS continuous at *every* $\boldsymbol{x}_0 \in \mathcal{X}$ provided $C_Y(\cdot)$ is continuous at the origin. Stated in terms of the correlation function, $C_Y(\boldsymbol{h}) \to C_Y(\boldsymbol{0}) = \sigma_Z^2$ as $\boldsymbol{h} \to \boldsymbol{0}$ is equivalent to

$$R_Y(\boldsymbol{h}) = C_Y(\boldsymbol{h})/\sigma_Z^2 \to 1.0 \quad \text{as} \quad \boldsymbol{h} \to \boldsymbol{0}.$$

Continuing our discussion of general random functions $Y(\cdot)$, Bochner (1955) proved that the covariance function of every stationary, MS continuous random function $Y(\cdot)$ on $\mathbb{R}^d$, can be written in the form

$$C_Y(\boldsymbol{h}) = \int_{\mathbb{R}^d} \cos(\boldsymbol{h}^\top \boldsymbol{w})\, dG(\boldsymbol{w}), \tag{2.3.9}$$

where $G(\cdot)$ is positive finite symmetric measure on $\mathbb{R}^d$. In particular, this characterization must hold for the special case of stationary GRFs. (See also the discussions in Cramér and Leadbetter (1967) on page 126, Adler (1981) on page 25, Cressie (1993) on page 84, or Stein (1999) on page 22-25.)

The process variance corresponding to $C_Y(\cdot)$ having the form (2.3.9) is

$$C_Y(\boldsymbol{0}) = \int_{\mathbb{R}^d} dG(\boldsymbol{w}) < +\infty$$

which is finite because G is a bounded measure on $\mathbb{R}^d$; $F(\cdot) = G(\cdot)/C_Y(\boldsymbol{0})$ is a symmetric probability distribution, called the *spectral distribution*, corresponding to $C_Y(\cdot)$. The function

$$R_Y(\boldsymbol{h}) = \int_{\mathbb{R}^d} \cos(\boldsymbol{h}^\top \boldsymbol{w})\, dF(\boldsymbol{w}) \tag{2.3.10}$$

is the correlation function corresponding to the spectral distribution $F(\cdot)$. If $F(\cdot)$ has a density $f(\cdot)$, then $f(\cdot)$ is called the *spectral density* corresponding to $R_Y(\cdot)$. In this case

$$R_Y(\boldsymbol{h}) = \int_{\mathbb{R}^d} \cos(\boldsymbol{h}^\top \boldsymbol{w}) f(\boldsymbol{w})\, d\boldsymbol{w}. \tag{2.3.11}$$

The right-hand side of (2.3.11) gives us a method to produce valid correlation functions (and covariance functions)—choose a symmetric density $f(\cdot)$ and evaluate the integral (2.3.11).

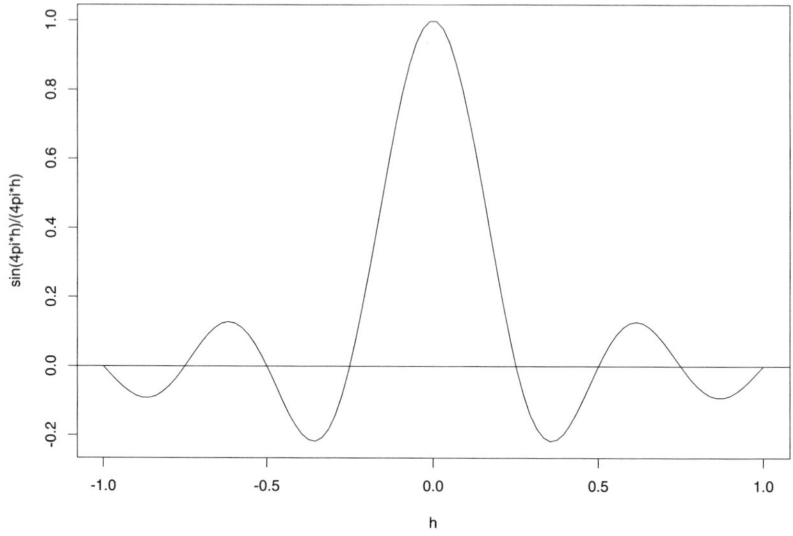

FIGURE 2.3. The correlation function $R(h) = \sin(h/\theta)/(h/\theta)$ for $\theta = 1/4\pi$ over h in $[-1, +1]$.

Example 2.2 This first example shows how (2.3.11) can be used to generate valid correlation functions from probability density functions that are symmetric about the origin. Consider the one-dimensional case. Perhaps the simplest choice of one-dimensional density is the uniform density over a symmetric interval which we take to be $(-1/\theta, +1/\theta)$ for a given $\theta > 0$. Thus the spectral density is

$$f(w) = \begin{cases} \theta/2, & -1/\theta < w < 1/\theta \\ 0, & \text{otherwise} \end{cases}$$

and the corresponding correlation function is

$$R(h) = \int_{-1/\theta}^{+1/\theta} \frac{\theta}{2} \cos(hw)\, dw = \begin{cases} \frac{\sin(h/\theta)}{h/\theta}, & h \neq 0 \\ 1, & h = 0 \end{cases}.$$

This correlation has scale parameter θ; Figure 2.3 shows that $R(h)$ can model both positive and negative correlations. ■

Any function $R_Y(\cdot)$ of the form (2.3.10) must satisfy $R_Y(\mathbf{0}) = 1$, must be continuous at $\boldsymbol{h} = \mathbf{0}$, must be symmetric about $\boldsymbol{h} = \mathbf{0}$, and must be positive semidefinite. The first consequence holds because

$$R_Y(\mathbf{0}) = \int_{IR^d} \cos(\mathbf{0}^\top \boldsymbol{w})\, dF(\boldsymbol{w}) = \int_{IR^d} 1\, dF(\boldsymbol{w}) = 1,$$

where the third equality in the above is true because $F(\cdot)$ is a probability distribution. Continuity follows by an application of the dominated convergence theorem; notice that from the argument following (2.3.8), continuity of $R_Y(\boldsymbol{h})$ at the origin insures that the corresponding process is MS continuous. Symmetry holds because $\cos(-x) = \cos(x)$ for all real x. Positive semidefinite is true because for any $L \geq 1$, any real numbers $w_1, \ldots, w_L$, and any $\boldsymbol{x}_1, \ldots, \boldsymbol{x}_L$ we have

$$
\begin{aligned}
&\sum_{i=1}^{L}\sum_{j=1}^{L} w_i w_j R_Y(\boldsymbol{x}_i - \boldsymbol{x}_j) \\
&= \int_{IR^d} \sum_{i=1}^{L}\sum_{j=1}^{L} w_i w_j \cos(\boldsymbol{x}_i^\top \boldsymbol{w} - \boldsymbol{x}_j^\top \boldsymbol{w})\, dF(\boldsymbol{w}) \\
&= \int_{IR^d} \sum_{i=1}^{L}\sum_{j=1}^{L} w_i w_j \left\{\cos(\boldsymbol{x}_i^\top \boldsymbol{w})\cos(\boldsymbol{x}_j^\top \boldsymbol{w}) \right. \\
&\qquad \left. + \sin(\boldsymbol{x}_i^\top \boldsymbol{w})\sin(\boldsymbol{x}_j^\top \boldsymbol{w})\right\}\, dF(\boldsymbol{w}) \\
&= \int_{IR^d} \left\{ \left(\sum_{i=1}^{L} w_i \cos(\boldsymbol{x}_i^\top \boldsymbol{w})\right)^2 + \left(\sum_{i=1}^{L} w_i \sin(\boldsymbol{x}_i^\top \boldsymbol{w})\right)^2 \right\} dF(\boldsymbol{w}) \\
&\geq 0.
\end{aligned}
$$

Continuity, symmetry, and positive semidefiniteness also hold for any covariance function $C_Y(\cdot)$ of form (2.3.9).

We conclude by mentioning several additional tools that are extremely useful for "building" covariance and correlation functions given a basic set of such functions. Suppose that $C_1(\cdot)$ and $C_2(\cdot)$ are valid covariance functions. Then their sum and product,

$$C_1(\cdot) + C_2(\cdot) \quad \text{and} \quad C_1(\cdot) \times C_2(\cdot),$$

are also valid covariance functions. The sum, $C_1(\cdot) + C_2(\cdot)$, is the covariance of two independent processes, one with covariance function $C_1(\cdot)$ and the other with covariance function $C_2(\cdot)$. Similarly, $C_1(\cdot) \times C_2(\cdot)$ is the covariance function of the product of two independent zero-mean GRFs with covariances $C_1(\cdot)$ and $C_2(\cdot)$, respectively.

The product of two valid correlation functions, $R_1(\cdot)$ and $R_2(\cdot)$, is a valid correlation function, but their sum is not (notice that $R_1(\boldsymbol{0}) + R_2(\boldsymbol{0}) = 2$, which is not possible for a correlation function). Correlation functions that are the products of one-dimensional marginal correlation functions are sometimes called *separable* correlation functions (not to be confused with the earlier use of the term separable).

We now introduce two widely–used families of correlation functions that have been used in the literature to specify stationary Gaussian stochastic processes (see also Journel and Huijbregts (1978), Mitchell, Morris and Ylvisaker (1990), Cressie (1993), Vecchia (1988), and Stein (1999)).

Example 2.3 Another familiar choice of a symmetric density that can be used as a spectral density is the normal density. To give a simple form for the resulting correlation function, take the spectral density to be $N(0, 2/\theta^2)$ for $\theta > 0$. Calculation gives

$$\begin{aligned} R(h) &= \int_{-\infty}^{+\infty} \cos(hw) \frac{\theta}{\sqrt{2\pi}\sqrt{2}} \exp\{-w^2\theta^2/4\}\, dw \\ &= \exp\left\{-(h/\theta)^2\right\}. \end{aligned} \tag{2.3.12}$$

This correlation is sometimes called the *Gaussian correlation function* because of its form but the reader should realize that the name is, perhaps, a misnomer. The Gaussian correlation function is a special case of the more general family of correlations called the power exponential correlation family. This family is far and away the most popular family of correlation models in the computer experiments literature. The one-dimensional GRF $Z(x)$ on $x \in \mathbb{R}$ has *power exponential* correlation function provided

$$R(h) = \exp\left\{-|h/\theta|^p\right\} \quad \text{for} \quad h \in \mathbb{R}, \tag{2.3.13}$$

where $\theta > 0$, and $0 < p \leq 2$. In addition to the Gaussian subfamily, the case $p = 1$,

$$R(h) = \exp\left\{-(|h|/\theta)\right\}$$

is well-studied. The GRF corresponding to this correlation function is known as the Ornstein-Uhlenbeck process.

For later reference, we note that every power exponential correlation function, $0 < p \leq 2$, is continuous at the origin, and none, except the Gaussian $p = 2$, is differentiable at the origin. In fact, the Gaussian correlation function is infinitely differentiable at the origin.

From the fact that products of correlation functions are also correlation functions,

$$R(\boldsymbol{h}) = \exp\left\{-\sum_{j=1}^{d} |h_j/\theta_j|^{p_j}\right\} \tag{2.3.14}$$

is a d-dimensional separable version of the power exponential correlation function, as is the special case of the product Gaussian family

$$R(\boldsymbol{h}) = \exp\left\{-\sum_{j=1}^{d} (h_j/\theta_j)^2\right\}$$

which has dimension–specific scale parameters. ■

Example 2.4 Suppose that $Z(x)$ is a one-dimensional GRF on $x \in \mathbb{R}$ with correlation function

$$R(h|\theta) = \begin{cases} 1 - 6\left(\frac{h}{\theta}\right)^2 + 6\left(\frac{|h|}{\theta}\right)^3, & |h| \leq \theta/2 \\ 2\left(1 - \frac{|h|}{\theta}\right)^3, & \theta/2 < |h| \leq \theta \\ 0, & \theta < |h| \end{cases}, \qquad (2.3.15)$$

where $0 < \theta$ and $h \in \mathbb{R}$. The function $R(h|\theta)$ has two continuous derivatives at $h = 0$ and also at the change point $h = \theta/2$ (see the right column of Figure 2.6). $R(h|\theta)$ assigns zero correlation to inputs x_1 and x_2 that are sufficiently far apart ($|x_1 - x_2| > \theta$). Formally, the spectral density that produces (2.3.15) is proportional to

$$\frac{1}{w^4\theta^3}\left\{72 - 96\cos(w\theta/2) + 24\cos(w\theta)\right\}.$$

Anticipating Section 3.2 on prediction in computer experiments, the use of (2.3.15) leads to cubic spline interpolating predictors. As in the previous example, we note that

$$R(\boldsymbol{h}|\boldsymbol{\theta}) = \prod_{j=1}^{d} R(h_j|\theta_j)$$

for $\boldsymbol{h} \in \mathbb{R}^d$ is a correlation function that allows each input dimension to have its own scale and thus dimension specific rate at which $Z(\cdot)$ values become uncorrelated. Other one-dimensional cubic correlation functions can be found in Mitchell et al. (1990) and Currin, Mitchell, Morris and Ylvisaker (1991). ■

2.3.4 Using the Correlation Function to Specify a GRF with Given Smoothness Properties

In practice we reduce the choice of a GRF to that of a covariance (or correlation) function whose realizations have desired prior smoothness characteristics. Hence we now turn attention to describing the relationship between the smoothness properties of a stationary GRF, $Z(\cdot)$, and the properties of its covariance function, $C(\cdot)$. To describe this relationship for general processes would require substantial space. By restricting attention to stationary GRFs we can provide a relatively concise overview. See Adler (1990), Abrahamsen (1997), or Stein (1999) for a discussion of these ideas for more general processes and for additional detail concerning the Gaussian process case.

There are several different types of "continuity" and "differentiability" that a process can possess. The definitions differ in their ease of application and the technical simplicity with which they are established. Given a particular property such as continuity at a point or differentiability over an interval, we would like to know that draws from a given random function model $Z(\cdot)$ have that property with probability one. For example, if Q is a property of interest, say continuity at the point $\boldsymbol{x}_0$, then we desire

$$P\{\omega : Z(\cdot, \omega) \text{ has property } Q\} = 1.$$

We term this *almost sure behavior* of the sample paths.

Subsection 2.3.3 introduced the widely-used concept of MS continuity. We saw an instance of the general fact that MS properties are relatively simple to prove, although they are not of direct interest in describing sample paths. Below we show that a slight strengthening of the conditions under which MS continuity holds guarantees almost sure continuity.

Recall that in Subsection 2.3.3 we stated that any stationary random function $Z(\cdot)$ on $\mathcal{X}$ having finite second moments is MS continuous on $\mathcal{X}$ provided that its correlation function is continuous at the origin, i.e., $R(\boldsymbol{h}) \to 1$ as $\boldsymbol{h} \to \boldsymbol{0}$. GRFs with either the cubic (2.3.15) or the power exponential (2.3.13) correlation functions are examples of such random functions.

Adler (1981) (page 60) shows that for the sample paths of stationary GRFs to be almost surely continuous, one need only add a condition requiring that $R(\boldsymbol{h})$ converge to unity sufficiently fast. For example, a consequence of his Theorem 3.4.1 is that, if $Z(\cdot)$ is a stationary GRF with correlation function $R(\cdot)$ that satisfies

$$1 - R(\boldsymbol{h}) \leq \frac{c}{|\log(\|\boldsymbol{h}\|_2)|^{1+\epsilon}} \quad \text{for all} \quad \|\boldsymbol{h}\|_2 < \delta \tag{2.3.16}$$

for some $c > 0$, some $\epsilon > 0$, and some $\delta < 1$, then $Z(\cdot)$ has almost surely continuous sample paths. MS continuity requires that $(1 - R(\boldsymbol{h})) \to 0$ as $\boldsymbol{h} \to 0$; the factor $|\log(\|\boldsymbol{h}\|_2)|^{1+\epsilon} \to +\infty$ as $\boldsymbol{h} \to 0$. Thus (2.3.16) holds provided that $1 - R(\boldsymbol{h})$ converges to zero at least as fast as $|\log(\|\boldsymbol{h}\|_2)|^{1+\epsilon}$ diverges to $+\infty$. The product

$$[1 - R(\boldsymbol{h})] \times \left|\log\left(\|\boldsymbol{h}\|_2\right)\right|^{1+\epsilon}$$

is bounded for most correlation functions used in practice. In particular this is true for any power exponential correlation function with $0 < p \leq 2$. One can also use the spectral distribution to give sufficient conditions for almost sure continuity of sample paths. The standard conditions are stated in terms of the finiteness of the moments of the spectral distribution. For example, see Theorem 3.4.3 of Adler (1981) or Sections 9.3 and 9.5 of Cramér and Leadbetter (1967).

Conditions for almost sure continuity of the sample paths of nonstationary GRFs, $Z(\cdot)$, can be similarly expressed in terms of the rate at which

$$E\left\{|Z(\boldsymbol{x}_1) - Z(\boldsymbol{x}_2)|^2\right\}$$

converges to zero as $\|\boldsymbol{x}_1 - \boldsymbol{x}_2\|_2 \to 0$ (Adler (1981), Theorem 3.4.1).

As for continuity, a concept of mean square differentiability can be defined that describes the mean difference of the usual tangent slopes of a given process and a limiting "derivative process." Instead, here we directly discuss the parallel to almost sure continuity. Consider the individual sample draws $z(\boldsymbol{x}) = Z(\boldsymbol{x}, \omega)$, $\mathcal{X} \subset \mathbb{R}^d$, corresponding to specific outcomes $\omega \in \Omega$. Suppose that the jth partial derivative of $Z(\boldsymbol{x}, \omega)$ exists for $j = 1, \ldots d$ and $\boldsymbol{x} \in \mathcal{X}$, i.e.,

$$\nabla_j Z(\boldsymbol{x}, \omega) = \lim_{\delta \to 0} \frac{Z(\boldsymbol{x} + \boldsymbol{e}_j \delta, \omega) - Z(\boldsymbol{x}, \omega)}{\delta}$$

exists where $\boldsymbol{e}_j$ denotes the unit vector in the jth direction. Let

$$\boldsymbol{\nabla} Z(\boldsymbol{x}, \omega) = (\nabla_1 Z(\boldsymbol{x}, \omega), \ldots, \nabla_d Z(\boldsymbol{x}, \omega))$$

denote the vector of partial derivatives of $Z(\boldsymbol{x}, \omega)$, sometimes called the gradient of $Z(\boldsymbol{x}, \omega)$. We will state conditions on the covariance (correlation) function that guarantee that the sample paths are almost surely differentiable. The situation for higher order derivatives can be described in a similar manner, sample pathwise, for each ω.

As motivation for the condition given below, we observe the following heuristic calculation that gives the covariance of the derivative of $Z(\cdot)$. Fix $\boldsymbol{x}_1$ and $\boldsymbol{x}_2$ in $\mathcal{X}$, then

$$\begin{aligned}
&\text{Cov}\Big(\tfrac{1}{\delta_1}(Z(\boldsymbol{x}_1 + \boldsymbol{e}_j\delta_1) - Z(\boldsymbol{x}_1))\ ,\ \tfrac{1}{\delta_2}(Z(\boldsymbol{x}_2 + \boldsymbol{e}_j\delta_2) - Z(\boldsymbol{x}_2))\Big) \\
&\quad = \frac{1}{\delta_1 \delta_2} \{C(\boldsymbol{x}_1 - \boldsymbol{x}_2 + \boldsymbol{e}_j(\delta_1 - \delta_2)) - C(\boldsymbol{x}_1 - \boldsymbol{x}_2 + \boldsymbol{e}_j\delta_1) \\
&\qquad - C(\boldsymbol{x}_1 - \boldsymbol{x}_2 - \boldsymbol{e}_j\delta_2) + C(\boldsymbol{x}_1 - \boldsymbol{x}_2)\} \\
&\quad \to - \left.\frac{\partial^2 C(\boldsymbol{h})}{\partial h_j^2}\right|_{\boldsymbol{h} = \boldsymbol{x}_1 - \boldsymbol{x}_2}
\end{aligned} \tag{2.3.17}$$

as $\delta_1, \delta_2 \to 0$ when the second partial derivative of $C(\cdot)$ exists. These calculations motivate the fact that the covariance function of the partial derivatives of $Z(\cdot)$, if they exist, are given by the partial derivatives of $C(\boldsymbol{h})$. Thus it should come as no surprise that to assure that a given Gaussian random field has, almost surely, differentiable draws, the conditions required are on the partial derivatives of the covariance function.

Formally, suppose

$$C_j^{(2)}(\boldsymbol{h}) \equiv \frac{\partial^2 C(\boldsymbol{h})}{\partial h_j^2}$$

exists and is continuous with $C_j^{(2)}(\boldsymbol{0}) \neq 0$; let $R_j^{(2)}(\boldsymbol{h}) \equiv C_j^{(2)}(\boldsymbol{h})/C_j^{(2)}(\boldsymbol{0})$ be the normalized version of $C_j^{(2)}(\cdot)$. Then almost surely $Z(\cdot)$ has j^{th} partial differentiable sample path, denoted $\nabla_j Z(\boldsymbol{x})$, provided $R_j^{(2)}(\cdot)$ satisfies

(2.3.16). In this case $-C_j^{(2)}(\boldsymbol{h})$ is the covariance function and $R_j^{(2)}(\boldsymbol{h})$ is the correlation function of $\nabla_j Z(\boldsymbol{x})$.

Higher order $Z(\cdot)$ derivatives can be iteratively developed in the same way, although a more sophisticated notation must be introduced to describe the higher-order partial derivatives required of $C(\cdot)$. Conditions for nonstationary $Z(\cdot)$ can be determined from almost sure continuity conditions for nonstationary $Z(\cdot)$ (Adler (1981), Chapter 3).

We complete this section by illustrating the effects of changing the covariance parameters on the draws of several stationary GRFs that were introduced earlier and on one important additional family, the Matérn correlation function. In each case, the plot was obtained by linearly joining draws from an appropriate 20 or 40 dimensional multivariate normal distribution; hence the figures give the spirit, if not the detail, of the sample paths from the associated process. The interested reader can gain additional feel for stationary Gaussian processes by using the software of Kozintsev (1999) or Kozintsev and Kedem (2000) for generating two-dimensional Gaussian random fields (see the URL

`http://www.math.umd.edu/~bnk/CLIP/clip.gauss.htm`

for details).

Example 2.3 (Continued–power exponential correlation function) Figures 2.4 and 2.5 show the marginal effects of changing the shape parameter p and the scale parameter θ on the function draws from GRFs over $[0, 1]$ having the power exponential correlation function (2.3.13). *These figures, and those that illustrate the other GRFs that are discussed below, connect 20 points drawn from a multivariate normal distribution having the desired covariance matrix and so illustrate the spirit of the function draws, if not their fine detail.*

For powers $p < 2$, the sample paths are theoretically nondifferentiable and this can be seen in the bottom two panels of Figure 2.4. The sample paths for $p = 2.0$ are infinitely differentiable; the draws in the top panel of Figure 2.4 are very near the process mean of zero for $\theta = 1.0$. As shown in Figure 2.5, the number of local maxima and minima in sample paths is controlled by the scale parameter when $p = 2.0$. Figure 2.5 shows that as the scale parameter θ *decreases*, the correlations for each fixed pair of inputs decreases and the sample paths have increasing numbers of local maxima. This is true because the process exhibits less dependence for "nearby" x and thus "wiggles" more like white noise, the case of uncorrelated $Z(\boldsymbol{x})$. As θ *increases*, the correlation for each pair of inputs increases and, as the correlation approaches unity, the draws become more nearly the constant zero, the process mean. In Figure 2.5 the most extreme case of this phenomenon is shown in the top panel where $(p, \theta) = (2.0, 0.50)$. ∎

Example 2.4 (Continued–cubic correlation function) Recall that the cubic correlation (and covariance) function (2.3.15) is twice continuously

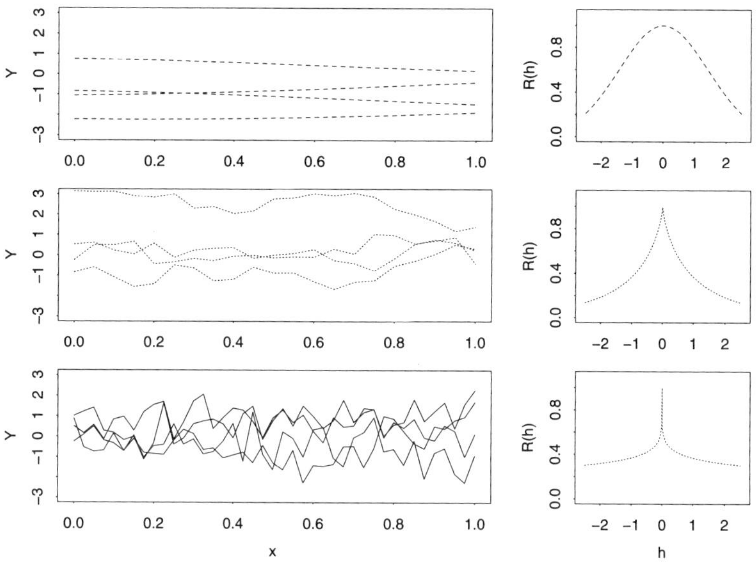

FIGURE 2.4. The Effect of Varying the Power on the Sample Paths of a GRF with a Power Exponential Correlation Function. Four draws from a zero mean, unit variance GRF with the exponential correlation (2.3.13) having fixed $\theta \equiv 1.0$ with $p = 2.0$ (dashed lines), $p = 0.75$ (dotted lines), and $p = 0.20$ (solid lines).

differentiable. Thus draws from a GRF with this correlation structure will be continuous and differentiable. Figure 2.6 shows draws from this process for different θ. As the scale parameter θ *decreases*, the domain where $R(h) = 0$ increases and hence the paths become more like white noise, i.e., having independent and identically distributed Gaussian components. As θ *increases*, the paths tend to become flatter with fewer local maxima and minima. ■

Example 2.5 The Matérn correlation function was introduced by Matérn in his thesis (Matérn (1960) or see the reprint Matérn (1986) and Vecchia (1988) for related work). This model has been used especially to describe the spatial and temporal variability in environmental data (see Rodríguez-Iturbe and Mejía (1974), Handcock and Stein (1993), Handcock and Wallis (1994), and especially Stein (1999)).

From the viewpoint of the spectral representation, the Matérn correlation function arises by choosing the t distribution as the spectral density. Given $\nu > 0$ and $\theta > 0$, use of the t density

$$f(w) = \frac{\Gamma(\nu + 1/2)}{\Gamma(\nu)\sqrt{\pi}} \left(\frac{4\nu}{\theta^2}\right)^{\nu} \frac{1}{\left(w^2 + \frac{4\nu}{\theta^2}\right)^{\nu+1/2}}$$

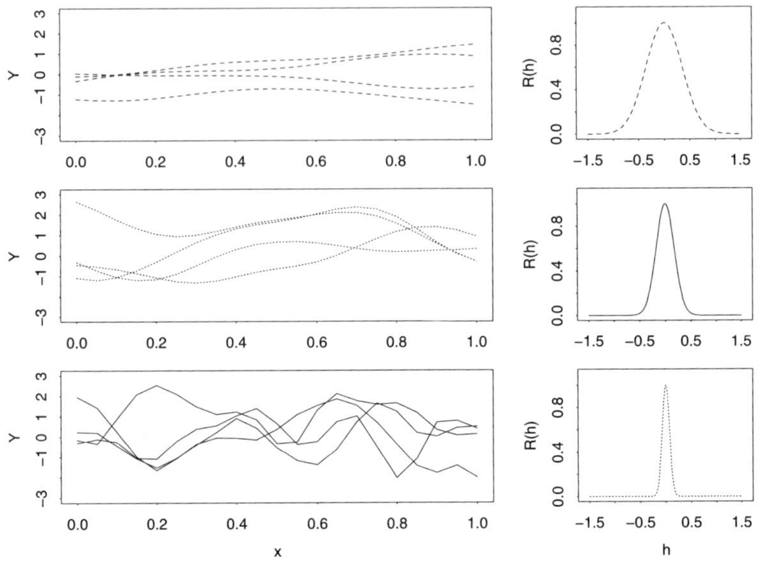

FIGURE 2.5. The Effect of Varying the Scale Parameter on the Sample Paths of a GRF with a Power Exponential Correlation Function. Four draws from a zero mean, unit variance GRF with the exponential correlation function (2.3.12) (having fixed $p = 2.0$) for $\theta = 0.50$ (dashed lines), $\theta = 0.25$ (dotted lines), and $\theta = 0.10$ (solid lines).

in spectral correlation formula (2.3.10) gives the two parameter correlation family

$$R(h) = \frac{1}{\Gamma(\nu)2^{\nu-1}} \left(\frac{2\sqrt{\nu}\,|h|}{\theta}\right)^{\nu} K_{\nu}\left(\frac{2\sqrt{\nu}\,|h|}{\theta}\right), \tag{2.3.18}$$

where $K_\nu(\cdot)$ is the modified Bessel function of order ν. As is usual in the literature, we refer to (2.3.18) as the Matérn correlation function. The parameter θ is clearly a scale parameter for this family. The modified Bessel function arises as the solution of a certain class of ordinary differential equations (Kreyszig (1999)). In general, $K_\nu(t)$ is defined in terms of an infinite power series in t; when ν equals a half integer, i.e., $\nu = n + 1/2$ for $n \in \{0, 1, 2, \ldots\}$, then $K_{n+1/2}(\cdot)$ can be expressed as the finite sum

$$K_{n+1/2}(t) = e^{-t}\sqrt{\frac{\pi}{2t}} \sum_{k=0}^{n} \frac{(n+k)!}{k!\,(n-k)!}\frac{1}{(2t)^k}.$$

The corresponding Matérn correlation function (2.3.18) is

$$e^{-2\sqrt{\nu}|h|/\theta}\left\{b_0\left(\frac{|h|}{\theta}\right)^{n} + b_1\left(\frac{|h|}{\theta}\right)^{n-1} + b_2\left(\frac{|h|}{\theta}\right)^{n-2} + \cdots + b_n\right\},$$

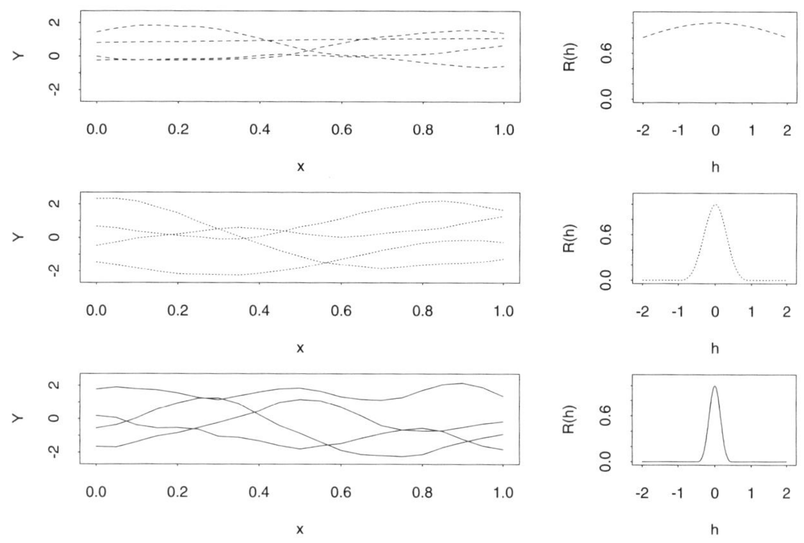

FIGURE 2.6. The Effect of Varying the Scale Parameter on the Sample Paths of a GRF with a Cubic Correlation Function. Four draws from a zero mean, unit variance GRF with the cubic correlation function (2.3.15) for $\theta = 0.5$ (solid lines), $\theta = 1.0$ (dotted lines), and $\theta = 10.0$ (dashed lines). The corresponding correlation function is plotted to the right of each set of sample paths.

where the coefficients are given by

$$b_j = \frac{\sqrt{\pi}\ \nu^{(n-j)/2}}{4^j \Gamma(\nu)} \ \frac{(n+j)!}{j!\ (n-j)!}$$

for $j = 0, 1, \ldots$ where $\nu = n + 1/2$; the b_j depend on ν but not θ. For example, when $n = 0$ ($\nu = 1/2$),

$$K_{1/2}(t) = \sqrt{\pi} e^{-t} / \sqrt{2t} \ \text{ and so } \ R(h) = e^{-\sqrt{2}|h|/\theta},$$

which is a special case of the power exponential correlation function with $p = 1$ that was introduced earlier. Similarly, $R(h) \to e^{-(h/\theta)^2}$ as $\nu \to \infty$ so that this class of correlations includes the Gaussian correlation function in the limit.

The smoothness of functions drawn from a GRF with Matérn correlation depends on ν. Let $\lceil \nu \rceil$ denote the integer ceiling of ν, i.e., the smallest integer that is greater than or equal to ν. For example, $\lceil 3.2 \rceil = 4$ and $\lceil 3 \rceil = 3$. Then functions drawn from a GRF having the Matérn correlation have almost surely continuously differentiable sample draws of order ($\lceil \nu \rceil - 1$). Thus we refer to ν as the smoothness parameter of the Matérn family (see Cramér and Leadbetter (1967)).

Products of the one-dimensional Matérn correlation function can be useful for modeling d-dimensional input responses. In this case, the family might include dimension specific scale parameters and a common smoothness parameter,

$$R(\boldsymbol{h}) = \prod_{i=1}^{d} \frac{1}{\Gamma(\nu)2^{\nu-1}} \left(\frac{2\sqrt{\nu}\,|h_i|}{\theta_i} \right)^{\nu} K_{\nu} \left(\frac{2\sqrt{\nu}\,|h_i|}{\theta_i} \right),$$

or dimension specific scale and smoothness parameters.

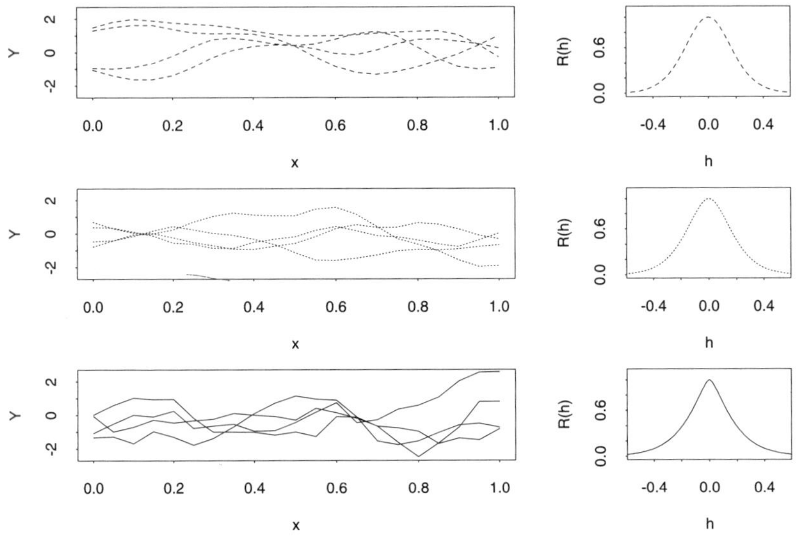

FIGURE 2.7. The Effect of Varying the ν Parameter on the Sample Paths of a GRF with Matérn Correlation Function. Four draws from a zero mean, unit variance GRF with the Matérn correlation function (2.3.18) (having fixed $\theta = 0.25$) for $\nu = 1$ (solid lines), $\nu = 2.5$ (dotted lines), and $\nu = 5$ (dashed lines).

We conclude by displaying sets of function draws from one-dimensional GRFs on $[0, 1]$ having different Matérn correlation functions to illustrate the effect of changing the scale and shape parameters.

Figure 2.7 fixes the scale parameter at $\theta = 0.25$ and varies $\nu \in \{1, 2.5, 5\}$. The draws clearly show the increase in smoothness as ν increases. As a practical matter, it is difficult for most observers to distinguish sample paths having 3 or 4 continuous derivatives from those that are infinitely differentiable. In contrast, Figure 2.8 fixes the smoothness parameter at $\nu = 4$ and varies $\theta \in \{0.01, 0.25, 2.0\}$. For fixed ν and $0 < h < 1.0$, the scaled range of $|h|/\theta$ varies substantially for different θ; $|h|/\theta$ ranges from

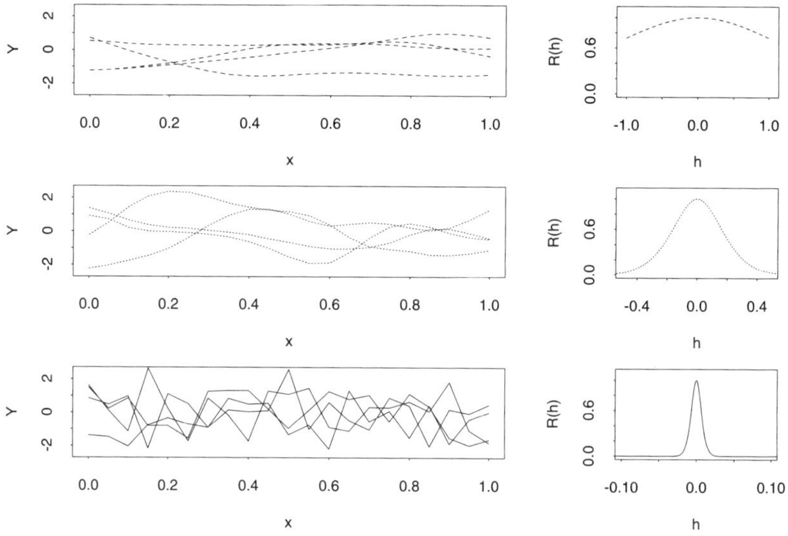

FIGURE 2.8. The Effect of Varying the Scale Parameter on the Sample Paths of a GRF with Matérn Correlation Function. Four draws from a zero mean, unit variance GRF with the Matérn correlation function (2.3.18) (having fixed $\nu = 4$) for $\theta = 0.01$ (solid lines), $\theta = 0.25$ (dotted lines), and $\theta = 2.0$ (dashed lines).

0.0 to 100 for $\theta = 0.01$ while this ratio only varies over 0.0 to 0.5 for $\theta = 2.0$. Notice that we use different h ranges for plotting $R(h)$ in Figure 2.8 to better illustrate the character of the correlation function near the origin. As θ decreases, the correlation function of any two fixed points decreases (to zero) and hence the sample paths "look" more like white noise. Thus the bottom panel of this figure plots a process with many more local maxima and minima than does the top panel. ■

2.3.5 *Hierarchical Gaussian Random Field Models*

While the examples above can provide guidance about the choice of a specific GRF prior for $y(\cdot)$, it will often be the case that the user will not be prepared to specify every detail of the GRF prior. For example, it will often be difficult to specify the correlation function of the GRF. A flexible alternative to the complete specification of a GRF is to use a *hierarchical* GRF prior model for $Y(\cdot)$. To describe this model, suppose that

$$Y(\boldsymbol{x}) = \sum_{j=1}^{p} f_j(\boldsymbol{x})\beta_j + Z(\boldsymbol{x}) = \boldsymbol{f}^\top(\boldsymbol{x})\boldsymbol{\beta} + Z(\boldsymbol{x}),$$

where $Z(\cdot)$ is a Gaussian random field with zero mean, variance σ_Z^2, and correlation function $R(\cdot \,|\, \boldsymbol{\psi})$. Here $R(\cdot \,|\, \boldsymbol{\psi})$ denotes a parametric family of correlation functions. In a hierarchical model some (or all) of $\boldsymbol{\beta}$, σ_Z^2, and $\boldsymbol{\psi}$ are not specified but rather a 2nd stage distribution that describes expert opinion about the relative likelihood of the parameter values is specified.

To be specific, suppose it desired to place a 2nd stage prior on all three parameters $\boldsymbol{\beta}$, σ_Z^2, and $\boldsymbol{\psi}$. Sometimes this task is facilitated because the prior $[\boldsymbol{\beta}, \sigma_Z^2, \boldsymbol{\psi}]$ can be expressed in "pieces." Suppose that it is reasonable to assume that large scale location parameters $\boldsymbol{\beta}$ and the small scale variance, σ_Z^2, are independent of the correlation parameters, $\boldsymbol{\psi}$. This means that

$$[\boldsymbol{\beta}, \sigma_Z^2, \boldsymbol{\psi}] = [\boldsymbol{\beta}, \sigma_Z^2] \times [\boldsymbol{\psi}] = [\boldsymbol{\beta} \,|\, \sigma_Z^2] \times [\sigma_Z^2] \times [\boldsymbol{\psi}] \; .$$

The second equality is true because $[\boldsymbol{\beta}, \sigma_Z^2] = [\boldsymbol{\beta} \,|\, \sigma_Z^2] \times [\sigma_Z^2]$ always holds. Thus the overall prior can be determined from these three pieces, which is often easier to do.

One complication with hierarchical models is that even when $[\boldsymbol{\beta}, \sigma_Z^2, \boldsymbol{\psi}]$ can be specified, it will usually be the case that the $Y(\boldsymbol{x})$ posterior cannot be expressed in closed form. Subsection 3.3.2 discusses the problem of computing the posterior mean in the context of various "empirical best linear unbiased predictors." See especially the discussion of "posterior mode empirical best linear unbiased predictors" beginning on page 68.

As an example, suppose that the input $\boldsymbol{x}$ is d-dimensional and that $R(\cdot \,|\, \boldsymbol{\psi})$ has the product Matérn correlation function

$$R(\boldsymbol{h} \,|\, \boldsymbol{\psi}) = \prod_{i=1}^{d} \frac{1}{\Gamma(\nu) 2^{\nu-1}} \left(\frac{2\sqrt{\nu}\,|h_i|}{\theta_i} \right)^{\nu} K_{\nu} \left(\frac{2\sqrt{\nu}\,|h_i|}{\theta_i} \right) \tag{2.3.19}$$

with unknown common smoothness parameter and dimension-specific scale parameters; thus $\boldsymbol{\psi} = (\theta_1, \ldots, \theta_d, \nu)$. Consider specification of prior $[\boldsymbol{\psi} = (\theta_1, \ldots, \theta_d, \nu)]$. Suppose that any ν, $2 \leq \nu \leq 50$ is equally likely, which implies that the number of derivatives in each dimension is equally likely to range from 1 to 49. Given ν, 2nd stage priors can be placed on each scale parameter by soliciting expert opinion about likelihood of correlation values between $Y(\boldsymbol{x}_1)$ and $Y(\boldsymbol{x}_2)$ where $\boldsymbol{x}_1$ and $\boldsymbol{x}_2$ differ in exactly one coordinate direction. See Oakley (2002) for details and a case study. There are other examples of the construction of 2nd stage prior distributions for parameters, mostly in the environmental literature. For example, Handcock and Wallis (1994) build a prior distribution for correlation parameters in their space-time model of the mean temperature of a region of the northern United States.

The references in the previous paragraph describe what might be thought of as "informative" 2nd stage priors. Again returning to the Matérn correlation function (2.3.19), it may be difficult to choose even the means and variances of the smoothness parameter and the scale parameters for specific dimensions, much less the $[\boldsymbol{\psi}]$ joint distribution. In such cases it is tempting to develop and use so-called "non-informative" 2nd stage priors, which give "equal" weight to all the legitimate parameter values. The reader should be warned that there is not always agreement in the statistical community about what constitutes a non-informative prior, even for parameters having finite ranges. Furthermore not every choice of a non-informative 2nd stage prior dovetails with the 1st stage model to produce a legitimate prior for $y(\cdot)$ (see the important paper by Berger, De Oliveira and Sansó (2001)). More will said about non-informative 2nd stage priors in Subsection 3.3.2 on page 68, which discusses "posterior mode empirical best linear unbiased predictors." Such predictors assume that a hierarchical GRF model is specified having parametric correlation function $R(\cdot \mid \boldsymbol{\psi})$ with unknown $\boldsymbol{\psi}$.

A third possible choice for a 2nd stage parameter prior is a "conjugate" prior. Conjugate priors lead to closed-form posterior calculations, and are sometimes reasonable. Subsection 4.1.2 discusses conjugate and non-informative 2nd stage $[\boldsymbol{\beta}]$ distributions (with σ_Z^2 and $\boldsymbol{\psi}$ known). Subsection 4.1.3 gives the analogous conjugate and non-informative 2nd stage $[\boldsymbol{\beta}, \sigma_Z^2]$ distributions (with $\boldsymbol{\psi}$ known). These two subsections give closed-form expressions for the posterior of $Y(\boldsymbol{x})$ given the data.

3

Predicting Output from Computer Experiments

3.1 Introduction

This chapter discusses techniques for predicting the output of a computer model based on training data. A naíve view of this problem might regard it as being *point estimation* of a *fixed population quantity.* In contrast, *prediction* is the problem of providing a point guess of the realization of a *random variable.* The reason why prediction is the relevant methodology for the computer experiment application will be discussed in Section 3.2.

Knowing how to predict computer output is a prerequisite for answering most practical research questions involving computer experiments. A partial list of such problems is given in Section 2.2. As an example, Section 6.3 will present a method for sequentially designing a computer experiment to find input conditions that maximize the computer output; this sequential design method uses the prediction methods developed in this chapter.

To introduce ideas, we initially consider the generic problem of predicting an arbitrary random variable Y_0 based on data $\boldsymbol{Y}^n = (Y_1, \ldots, Y_n)^\top$. When Y_0 and $\boldsymbol{Y}^n$ are dependent random quantities, it seems intuitive that $\boldsymbol{Y}^n$ contains information about Y_0. Hence it should be no surprise that the Y_0 predictors we introduce in this chapter depend on the joint distribution of Y_0 and $\boldsymbol{Y}^n$. Section 3.2 describes several Y_0 predictors based on various optimality and convenience criteria.

Section 3.3 applies these methods to predict $Y(\boldsymbol{x}_0)$, the computer output at $\boldsymbol{x}_0$, based on training data $(\boldsymbol{x}_i, Y(\boldsymbol{x}_i))$, $1 \leq i \leq n$. The reader who is familiar with regression methodology might think of using a flexible regres-

Definition The predictor $\widehat{Y}_0$ of Y_0 is a *minimum MSPE predictor* at F provided

$$\text{MSPE}(\widehat{Y}_0, F) \le \text{MSPE}(Y_0^\star, F) \tag{3.2.6}$$

for any alternative predictor $Y_0^\star$.

Minimum MSPE predictors are also called *best MSPE predictors.* Predictors of practical importance will simultaneously minimize the MSPE for *many* distributions F.

The fundamental theorem of prediction shows that the conditional mean of Y_0 given $\boldsymbol{Y}^n$ is the minimum MSPE predictor of Y_0 based on $\boldsymbol{Y}^n$.

Theorem 3.2.1 Suppose that $(Y_0, \boldsymbol{Y}^n)$ has a joint distribution F for which the conditional mean of Y_0 given $\boldsymbol{Y}^n$ exists. Then

$$\widehat{Y}_0 = E\{Y_0 \,|\, \boldsymbol{Y}^n\}$$

is the best MSPE predictor of Y_0.

Proof: Fix an arbitrary unbiased predictor $Y_0^\star = Y_0^\star(\boldsymbol{Y}^n)$,

$$\begin{aligned}
\text{MSPE}(Y_0^\star, F) &= E_F\{(Y_0^\star - Y_0)^2\} \\
&= E_F\left\{(Y_0^\star - \widehat{Y}_0 + \widehat{Y}_0 - Y_0)^2\right\} \\
&= E_F\left\{(Y_0^\star - \widehat{Y}_0)^2\right\} + \text{MSPE}(\widehat{Y}_0, F) \\
&\quad + \ 2E_F\left\{(Y_0^\star - \widehat{Y}_0)(\widehat{Y}_0 - Y_0)\right\} \\
&\ge \text{MSPE}\left(\widehat{Y}_0, F\right) \\
&\quad + \ 2E_F\left\{\left(Y_0^\star - \widehat{Y}_0\right)\left(\widehat{Y}_0 - Y_0\right)\right\} \qquad (3.2.7) \\
&= \text{MSPE}\left(\widehat{Y}_0, F\right),
\end{aligned}$$

where the final equality holds because

$$\begin{aligned}
E_F\left\{\left(Y_0^\star - \widehat{Y}_0\right)\left(\widehat{Y}_0 - Y_0\right)\right\} &= E_F\left\{\left(Y_0^\star - \widehat{Y}_0\right) E_F\left\{\left(\widehat{Y}_0 - Y_0\right) \,|\, \boldsymbol{Y}^n\right\}\right\} \\
&= E_F\left\{\left(Y_0^\star - \widehat{Y}_0\right)\left(\widehat{Y}_0 - E_F\left\{Y_0 \,|\, \boldsymbol{Y}^n\right\}\right)\right\} \\
&= E_F\left\{\left(Y_0^\star - \widehat{Y}_0\right) \times 0\right\} \\
&= 0. \ \square
\end{aligned}$$

There are two interesting properties of the best MSPE predictor that can be seen from the proof of Theorem 3.2.1. The first is that the conditional mean $\widehat{Y}_0$ is essentially the unique best MSPE predictor in many cases that arise in practice. This is because $\text{MSPE}(\widehat{Y}_0, F)$ and $\text{MSPE}(Y_0^\star, F)$ are equal if and only if equality holds in (3.2.7), which occurs when $\widehat{Y}_0 = Y_0^\star$ almost everywhere. The second is that $\widehat{Y}_0 = E\{Y_0 \mid \boldsymbol{Y}^n\}$ must be *unbiased* with respect to the model F for $(Y_0, \boldsymbol{Y}^n)$ because

$$E\{\widehat{Y}_0\} = E\{E\{Y_0 \mid \boldsymbol{Y}^n\}\} = E\{Y_0\}.$$

Example 3.1 (Continued–best MSPE predictors) Consider finding the minimum MSPE predictor $\widehat{Y}_0 = E\{Y_0 \mid \boldsymbol{Y}^n\}$ when the components of $(Y_0, \boldsymbol{Y}^n)$ are not merely uncorrelated but are independent $N(\beta_0, \sigma_\epsilon^2)$ random variables. By the independence of $Y_0, Y_1, \ldots, Y_n$, the conditional distribution $[Y_0 \mid \boldsymbol{Y}^n]$ is simply the $N(\beta_0, \sigma_\epsilon^2)$ marginal distribution of Y_0. In particular,

$$\widehat{Y}_0 = E\{Y_0 \mid \boldsymbol{Y}^n\} = \beta_0$$

is the best MSPE predictor. Notice that this minimum MSPE predictor changes with β_0 and thus is specific to this particular $(Y_0, \boldsymbol{Y}^n)$ joint distribution.

Now consider a more interesting two-stage model for the distribution of $(Y_0, \boldsymbol{Y}^n)$. Assume that σ_ϵ^2 is known and that the distribution specified in the previous paragraph is the first-stage (conditional) distribution of $(Y_0, \boldsymbol{Y}^n)$ given β_0, denoted by $[Y_0, \boldsymbol{Y}^n \mid \beta_0]$. Combine this first-stage distribution with the non-informative second-stage distribution

$$[\beta_0] \propto 1$$

for β_0. While improper priors need not produce proper posterior distributions, in this case one can show

$$[Y_0, \boldsymbol{Y}^n] = \int [Y_0, \boldsymbol{Y}^n \mid \beta_0]\, [\beta_0]\, d\beta_0$$

gives a *proper* joint distribution of $(Y_0, \boldsymbol{Y}^n)$. Using this $(Y_0, \boldsymbol{Y}^n)$ distribution, the conditional distribution

$$[Y_0 \mid \boldsymbol{Y}^n = \boldsymbol{y}^n] \sim N_1 \left[\overline{\boldsymbol{y}}_n, \sigma_\epsilon^2 \left(1 + \frac{1}{n}\right)\right]$$

can be calculated where $\overline{\boldsymbol{y}}_n = (\sum_{i=1}^n y_i)/n$ is the sample mean of the training data. It follows that, for this two-stage model, the minimum MSPE predictor of Y_0 is $\widehat{Y}_0 = (\sum_{i=1}^n Y_i)/n$. ■

Example 3.2 (More best MSPE predictors) Consider the regression model developed in Chapter 2 in which

$$Y_i \equiv Y(\boldsymbol{x}_i) = \sum_{j=1}^{p} f_j(\boldsymbol{x}_i)\beta_j + Z(\boldsymbol{x}_i) = \boldsymbol{f}^\top(\boldsymbol{x}_i)\boldsymbol{\beta} + Z(\boldsymbol{x}_i) \qquad (3.2.8)$$

for $0 \le i \le n$, where the $\{f_j(\cdot)\}$ are known regression functions, $\boldsymbol{\beta}$ is a *given* nonzero $p \times 1$ vector, and $Z(\boldsymbol{x})$ is a zero mean stationary Gaussian process with dependence specified by the covariance

$$\text{Cov}\{Z(\boldsymbol{x}_i), Z(\boldsymbol{x}_j)\} = \sigma_Z^2 R(\boldsymbol{x}_i - \boldsymbol{x}_j)$$

for some *known* correlation function $R(\cdot)$ (see Section 2.2). Then the joint distribution of $Y_0 = Y(\boldsymbol{x}_0)$ and $\boldsymbol{Y}^n = (Y(\boldsymbol{x}_1), \ldots, Y(\boldsymbol{x}_n))^\top$ is the multivariate normal distribution

$$\begin{pmatrix} Y_0 \\ \boldsymbol{Y}^n \end{pmatrix} \sim N_{1+n}\left[\begin{pmatrix} \boldsymbol{f}_0^\top \\ \boldsymbol{F} \end{pmatrix}\boldsymbol{\beta}, \sigma_Z^2 \begin{pmatrix} 1 & \boldsymbol{r}_0^\top \\ \boldsymbol{r}_0 & \boldsymbol{R} \end{pmatrix}\right], \tag{3.2.9}$$

where $\boldsymbol{f}_0 = \boldsymbol{f}(\boldsymbol{x}_0)$ is the $p \times 1$ vector of regressors at $\boldsymbol{x}_0$, $\boldsymbol{F}$ is the $n \times p$ matrix of regressors having $(i,j)th$ element $f_j(\boldsymbol{x}_i)$ for $1 \le i \le n, 1 \le j \le p$, $\boldsymbol{\beta}$ is a $p \times 1$ vector of unknown regression coefficients, and the $n \times 1$ vector $\boldsymbol{r}_0 = (R(\boldsymbol{x}_0 - \boldsymbol{x}_1), \ldots, R(\boldsymbol{x}_0 - \boldsymbol{x}_n))^\top$ and $n \times n$ matrix $\boldsymbol{R} = (R(\boldsymbol{x}_i - \boldsymbol{x}_j))$ are defined in terms of the correlation function $R(\cdot)$. Assuming that the design matrix $\boldsymbol{F}$ is of full column rank p and that $\boldsymbol{R}$ is positive definite, Theorems 3.2.1 and B.1.2 show that

$$\widehat{Y}_0 = E\{Y_0 \,|\, \boldsymbol{Y}^n\} = \boldsymbol{f}_0^\top \boldsymbol{\beta} + \boldsymbol{r}_0^\top \boldsymbol{R}^{-1}(\boldsymbol{Y}^n - \boldsymbol{F}\boldsymbol{\beta}) \tag{3.2.10}$$

is the best MSPE predictor of Y_0.

The class of distributions $\mathcal{F}$ for which (3.2.10) is the minimum MSPE predictor is again embarrassingly small. In addition to $\widehat{Y}_0$ depending on the multivariate normality of $(Y_0, \boldsymbol{Y}^n)$, it also depends on *both* $\boldsymbol{\beta}$ and the specific correlation function $R(\cdot)$. Thus the best MSPE predictor changes when either $\boldsymbol{\beta}$ or $R(\cdot)$ changes, however, $\widehat{Y}_0$ is the same for all $\sigma_Z^2 > 0$.

As a final illustration, consider finding the minimum MSPE predictor of $Y(\boldsymbol{x}_0)$ based the following two-stage model for the regression data $(\boldsymbol{x}_i, Y(\boldsymbol{x}_i))$, $0 \le i \le n$. Suppose that (3.2.9) specifies the conditional distribution of $(Y_0, \boldsymbol{Y}^n)$ given $\boldsymbol{\beta}$ as the first stage of a two-stage model. (Assuming σ_Z^2 is known, say, although this is not needed). The second stage of the model puts an arbitrary prior on $(\boldsymbol{\beta}, \sigma_Z^2)$. The best MSPE predictor of Y_0 is

$$\begin{aligned}\widehat{Y_0} = E\{Y_0 \,|\, \boldsymbol{Y}^n\} &= E\{E\{Y_0 \,|\, \boldsymbol{Y}^n, \boldsymbol{\beta}\} \,|\, \boldsymbol{Y}^n\} \\ &= E\left\{\boldsymbol{f}_0^\top \boldsymbol{\beta} + \boldsymbol{r}_0^\top \boldsymbol{R}^{-1}(\boldsymbol{Y}^n - \boldsymbol{F}\boldsymbol{\beta}) \,\middle|\, \boldsymbol{Y}^n\right\}\end{aligned}$$

and the last expectation is with respect to the conditional distribution of $\boldsymbol{\beta}$ given $\boldsymbol{Y}^n$. Thus

$$\widehat{Y_0} = \boldsymbol{f}_0^\top E\{\boldsymbol{\beta} \,|\, \boldsymbol{Y}^n\} + \boldsymbol{r}_0^\top \boldsymbol{R}^{-1}(\boldsymbol{Y}^n - \boldsymbol{F} E\{\boldsymbol{\beta}|\boldsymbol{Y}^n\}) \tag{3.2.11}$$

is the minimum MSPE of Y_0 for *any* two-stage model whose first stage is given by (3.2.9) and has arbitrary second stage $\boldsymbol{\beta}$ prior for which $E\{\boldsymbol{\beta}|\boldsymbol{Y}^n\}$ exists.

Of course, the explicit formula for $E\{\boldsymbol{\beta}|\boldsymbol{Y}^n\}$, and hence $\widehat{Y_0}$, depends on the $\boldsymbol{\beta}$ prior. For example, when $\boldsymbol{\beta}$ has the non-informative prior, $[\boldsymbol{\beta}] \propto 1$, the conditional distribution $[\boldsymbol{\beta}|\boldsymbol{Y}^n]$ can be derived by observing

$$\begin{aligned}
[\boldsymbol{\beta}|\boldsymbol{Y}^n = \boldsymbol{y}^n] &\propto [\boldsymbol{y}^n|\boldsymbol{\beta}]\,[\boldsymbol{\beta}] \\
&\propto \exp\left\{-\frac{1}{2\sigma_Z^2}\left(\boldsymbol{y}^n - \boldsymbol{F}\boldsymbol{\beta}\right)^\top \boldsymbol{R}^{-1}\left(\boldsymbol{y}^n - \boldsymbol{F}\boldsymbol{\beta}\right)\right\} \times 1 \\
&\propto \exp\left\{-\frac{1}{2\sigma_Z^2}\left(\boldsymbol{\beta}^\top \boldsymbol{F}^\top \boldsymbol{R}^{-1}\boldsymbol{F}\boldsymbol{\beta} - 2\boldsymbol{\beta}^\top \boldsymbol{F}^\top \boldsymbol{R}^{-1}\boldsymbol{y}^n\right)\right\} \\
&= \exp\left\{-\frac{1}{2}\boldsymbol{\beta}^\top \boldsymbol{A}^{-1}\boldsymbol{\beta} + \boldsymbol{\nu}^\top \boldsymbol{\beta}\right\}, \text{ say,}
\end{aligned}$$

where $\boldsymbol{A}^{-1} = \boldsymbol{F}^\top(\sigma_Z^2\boldsymbol{R})^{-1}\boldsymbol{F}$ and $\boldsymbol{\nu} = \boldsymbol{F}^\top(\sigma_Z^2\boldsymbol{R})^{-1}\boldsymbol{y}^n$. Notice that $\text{rank}(\boldsymbol{A}) = p$ under the continuing assumption that $\boldsymbol{F}$ has full column rank p. Applying (B.1.2) of Appendix B gives

$$[\boldsymbol{\beta}|\boldsymbol{Y}^n] \sim N_p\left[(\boldsymbol{F}^\top \boldsymbol{R}^{-1}\boldsymbol{F})^{-1}\boldsymbol{F}^\top \boldsymbol{R}^{-1}\boldsymbol{Y}^n, \sigma_Z^2(\boldsymbol{F}^\top \boldsymbol{R}^{-1}\boldsymbol{F})^{-1}\right]$$

because the σ_Z^2 terms cancel in the expression for the mean of $\boldsymbol{\beta}|\boldsymbol{Y}^n$. Thus the best MSPE predictor of Y_0 under this two-stage model is

$$\widehat{Y_0} = \boldsymbol{f}_0^\top\,\widehat{\boldsymbol{\beta}} + \boldsymbol{r}_0^\top \boldsymbol{R}^{-1}\left(\boldsymbol{Y}^n - \boldsymbol{F}\widehat{\boldsymbol{\beta}}\right), \tag{3.2.12}$$

where $\widehat{\boldsymbol{\beta}} = (\boldsymbol{F}^\top \boldsymbol{R}^{-1}\boldsymbol{F})^{-1}\boldsymbol{F}^\top \boldsymbol{R}^{-1}\boldsymbol{Y}^n$. ■

There are at least three useful ways of thinking about the predictor (3.2.12). The first way is to regard (3.2.12) as the sum of the regression predictor of $\boldsymbol{f}_0^\top\widehat{\boldsymbol{\beta}}$ plus the "correction" $\boldsymbol{r}_0^\top \boldsymbol{R}^{-1}\left(\boldsymbol{Y}^n - \boldsymbol{F}\widehat{\boldsymbol{\beta}}\right)$. The second way of viewing (3.2.12) is as a function of the training data $\boldsymbol{Y}^n$; this viewpoint is important for describing the statistical properties of $\widehat{Y_0}$. The final method of examining formula (3.2.12) is as a function of $\boldsymbol{x}_0$, the point at which the prediction is to be made. The remainder of this subsection and Example 3.3 considers the nature of the correction in $\widehat{Y_0}$. We will return to the latter two methods of thinking about $\widehat{Y_0}$ in Section 3.3.

The correction term in (3.2.12) is $\boldsymbol{r}_0^\top \boldsymbol{R}^{-1}\left(\boldsymbol{Y}^n - \boldsymbol{F}\widehat{\boldsymbol{\beta}}\right)$, which is a linear combination of the residuals $\boldsymbol{Y}^n - \boldsymbol{F}\widehat{\boldsymbol{\beta}}$ based on the model (3.2.8) with prediction point specific coefficients, i.e.,

$$\boldsymbol{r}_0^\top \boldsymbol{R}^{-1}\left(\boldsymbol{Y}^n - \boldsymbol{F}\widehat{\boldsymbol{\beta}}\right) = \sum_{i=1}^{n} c_i(\boldsymbol{x}_0)\left(\boldsymbol{Y}^n - \boldsymbol{F}\widehat{\boldsymbol{\beta}}\right)_i, \tag{3.2.13}$$

where the weight $c_i(\boldsymbol{x}_0)$ is the ith element of $\boldsymbol{R}^{-1}\boldsymbol{r}_0$ and $\left(\boldsymbol{Y}^n - \boldsymbol{F}\widehat{\boldsymbol{\beta}}\right)_i$ is the ith residual based on the fitted model.

Example 3.3 To illustrate the regression and correction terms in the predictor (3.2.12), suppose the true unknown curve is the one-dimensional dampened cosine function

$$y(x) = e^{-1.4x} \cos(7\pi x/2)$$

over $0 \le x \le 1$ (see the top panel in Figure 3.1). We use a seven point training data set (also shown in Figure 3.1). The training data locations x_i were determined by selecting x_1 at random in the interval $[0,\, 1/7]$ and then adding $i/7$ to x_1 for $1 \le i \le 6$ to obtain six additional points. These next six x_i are equally spaced and located in the intervals $[1/7,\, 2/7]$, ..., $[6/7,\, 1]$. The choice of a design of the computer experiment will be discussed in Chapters 5 and 6.

Consider prediction of $y(\cdot)$ based on the stationary stochastic Gaussian process

$$Y(x) = \beta_0 + Z(x),$$

where $Z(\cdot)$ has zero mean, variance σ_Z^2, (Gaussian) correlation function

$$R(h) = e^{-136.1 \times h^2} .$$

Here $\boldsymbol{F} = \boldsymbol{1}_7$ is a 7×1 column vector of ones and, by (3.2.13), the predictor (3.2.12) is

$$\widehat{Y}(x_0) = \widehat{\beta_0} + \sum_{i=1}^{7} c_i(x_0) \left(Y_i - \widehat{\beta_0} \right)$$

when viewed as a function of x_0 where $\{x_i\}_{i=1}^7$ are the training data and $(Y_i - \widehat{\beta_0})$ is the ith residual from fitting the constant model. In this case, the regression predictor is $\widehat{\beta_0}$.

Consider specifically the prediction of $y(x_0)$ at $x_0 = 0.55$. The seven residuals $Y_i - \widehat{\beta_0}$ and their associated weights $c_i(0.55)$, $1 \le i \le 7$, are plotted in Figure 3.1. Notice that (1) the weights can be positive or negative and (2) the correction to the regression $\widehat{\beta_0}$ is based primarily on the residuals at the two training sites nearest to $x_0 = 0.55$; in fact, the three weights for the three training data points that are furthest from $x_0 = 0.55$ are indistinguishable from zero. ■

Returning to the general discussion of the correction $\boldsymbol{r}_0^\top \boldsymbol{R}^{-1}(\boldsymbol{Y}^n - \boldsymbol{F}\widehat{\boldsymbol{\beta}})$, we show that this term forces the predictor to *interpolate* the training data. To see why this is the case, suppose that $\boldsymbol{x}_0 = \boldsymbol{x}_i$ for some fixed i, $1 \le i \le n$. Then $\boldsymbol{f}_0 = \boldsymbol{f}^\top(\boldsymbol{x}_i)$ and

$$\boldsymbol{r}_0^\top = (\, R(\boldsymbol{x}_i - \boldsymbol{x}_1), R(\boldsymbol{x}_i - \boldsymbol{x}_2), \ldots, R(\boldsymbol{x}_i - \boldsymbol{x}_n)\,)$$

which is the ith row of $\boldsymbol{R}$. Thus $\boldsymbol{R}^{-1}\boldsymbol{r}_0 = (0, \ldots, 0, 1, 0, \ldots, 0)^\top = \boldsymbol{e}_i$, the ith unit vector, because this product is the ith column of $\boldsymbol{R}^{-1}\boldsymbol{R} = \boldsymbol{I}_n$, the

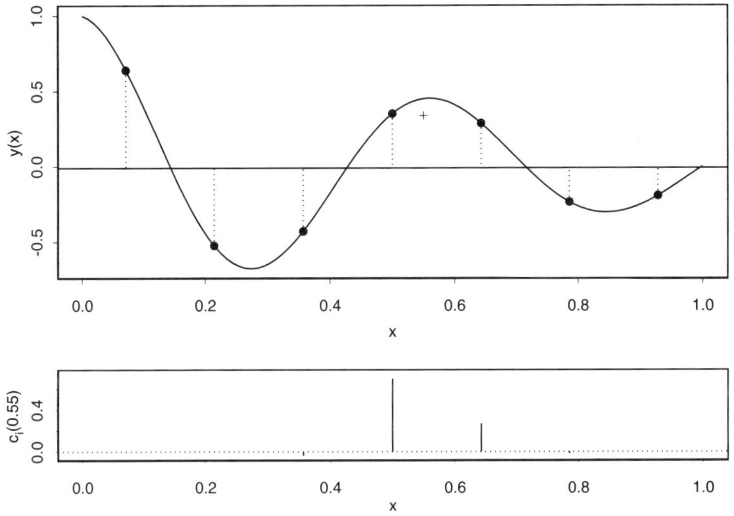

FIGURE 3.1. The top panel shows the true curve $y(x) = \exp\{-1.4x\} \times \cos(3.5\pi x)$ (solid line); the seven point input training data (dots); the BLUP at $x_0 = 0.55$ (cross); and the residuals, $Y_i - \widehat{\beta}_0$, (vertical dotted lines). The bottom panel plots the *weight* at each training data point as a line segment of length $|c_i(0.55)|$ from the origin with negative $c_i(0.55)$ plotted downward and positive $c_i(0.55)$ plotted upward.

$n \times n$ identity matrix. Hence

$$\boldsymbol{r}_0^\top \boldsymbol{R}^{-1} \left(\boldsymbol{Y}^n - \boldsymbol{F}\widehat{\boldsymbol{\beta}} \right) = \boldsymbol{e}_i^\top \left(\boldsymbol{Y}^n - \boldsymbol{F}\widehat{\boldsymbol{\beta}} \right) = Y_i - \boldsymbol{f}^\top(\boldsymbol{x}_i)\widehat{\boldsymbol{\beta}}$$

and so

$$\widehat{Y}(\boldsymbol{x}_0) = \boldsymbol{f}^\top(\boldsymbol{x}_i)\widehat{\boldsymbol{\beta}} + \left(Y_i - \boldsymbol{f}^\top(\boldsymbol{x}_i)\widehat{\boldsymbol{\beta}} \right) = Y_i \,.$$

Although we focus on the case of nonzero dependence in this book, we note that the argument above shows that for regression data with white noise (independent) measurement errors added to the mean of each observation, i.e., when the $Y_1, \ldots, Y_n$ are independent with Y_i having mean $\boldsymbol{f}^\top(\boldsymbol{x}_i)\boldsymbol{\beta}$, then $\boldsymbol{r}_0 = (0, \ldots, 0)^\top$. In this case the best MSPE predictor, expression (3.2.12), reduces to $\widehat{Y}(\boldsymbol{x}_0) = \boldsymbol{f}^\top(\boldsymbol{x}_0)\widehat{\boldsymbol{\beta}}$ for $\boldsymbol{x}_0 \neq \boldsymbol{x}_i$ where $\widehat{\boldsymbol{\beta}}$ is the ordinary least squares estimator of the mean of $Y(\boldsymbol{x}_0)$ and $\widehat{Y}(\boldsymbol{x}_0) = Y_i$ for $\boldsymbol{x}_0 = \boldsymbol{x}_i$. Thus the best MSPE predictor *interpolates* the data but has discontinuities at each of the data points.

Expression (3.2.12) is the basis for most predictors used in computer experiments. The next subsection shows that (3.2.12) has additional optimality properties that help explain its popularity. Before beginning this topic, we present a final example to show that best MSPE predictors need *not* be a linear function of the training data lest the previous (Gaussian model) examples, where the predictors are all linear in the data, suggest otherwise to the reader.

Example 3.4 Suppose that (Y_0, Y_1) has the joint distribution given by the density

$$f(y_0, y_1) = \begin{cases} 1/y_1^2, & 0 < y_1 < 1, 0 < y_0 < y_1^2 \\ 0, & \text{otherwise.} \end{cases}$$

Then it is straightforward to calculate that the conditional distribution of Y_0 given $Y_1 = y_1$ is uniform over the interval $(0, y_1^2)$. Hence the best MSPE predictor of Y_0 is the center of this interval, i.e., $\widehat{Y}_0 = E\{Y_0|Y_1\} = Y_1^2/2$ which is nonlinear in Y_1. In contrast, the minimum MSPE *linear unbiased* predictor of Y_0 is that $a_0 + a_1 Y_1$ which minimizes $E\{(a_0 + a_1 Y_1 - Y_0)^2\}$ among those (a_0, a_1) that satisfy the unbiasedness requirement $E\{a_0 + a_1 Y_1\} = E\{Y_0\}$. Unbiasedness leads to the restriction

$$a_0 + a_1 \frac{1}{2} = \frac{1}{6} \quad \text{or} \quad a_0 = \frac{1}{6} - a_1 \frac{1}{2}.$$

Applying calculus to minimize the MSPE

$$E\left\{\left(\left(\frac{1}{6} - a_1 \frac{1}{2}\right) + a_1 Y_1 - Y_0\right)^2\right\}$$

(expressed in terms of a_1) shows that $a_1 = 1/2$ (and $a_0 = 1/6 - a_1/2 = -1/12$), i.e., $\widehat{Y}_0^L = -\frac{1}{12} + \frac{1}{2} Y_1$ is the minimum MSPE linear unbiased predictor of Y_0.

As Figure 3.2 shows, the predictors $\widehat{Y}_0$ and $\widehat{Y}_0^L$ are very close over their $(0, 1)$ domain. The MSPE of $\widehat{Y}_0$ is obtained from

$$\begin{aligned} E\left\{\left(Y_0 - Y_1^2/2\right)^2\right\} &= E\left\{E\left\{(Y_0 - Y_1^2/2)^2 \middle| Y_1\right\}\right\} \\ &= E\left\{\text{Var}\{Y_0|Y_1\}\right\} \\ &= E\left\{Y_1^2/12\right\} \qquad (3.2.14) \\ &= 1/60 \approxeq 0.01667. \end{aligned}$$

The inner term $Y_1^2/12$ in (3.2.14) is the variance of the uniform distribution over $(0, y_1^2)$. A similar calculation gives the MSPE of $\widehat{Y}_0^L$ to be 0.01806 which is greater than the MSPE of the unconstrained predictor, as theory dictates, but the difference is small, as Figure 3.2 suggests. ■

3.2.3 Best Linear Unbiased MSPE Predictors

As we have seen, minimum MSPE predictors depend in detail on the joint distribution of the training data and Y_0; this criterion typically leads to

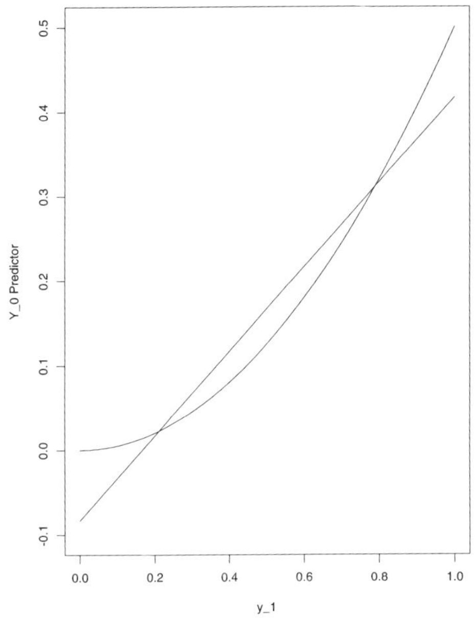

FIGURE 3.2. The predictors $\widehat{Y}_0$ and $\widehat{Y}_0^L$ based on $y_1 \in (0, 1)$

optimality within a very restricted class of competing predictors. In an attempt to find predictors that are optimal for a broader class of models, we focus on the two simpler types of predictors that were introduced in Example 3.4. Firstly we consider the class of Y_0 predictors that are linear in $\boldsymbol{Y}^n$, and secondly the class of predictors that are *both* linear and unbiased for Y_0.

The predictor $\widehat{Y}_0$ is a *minimum MSPE linear predictor* of Y_0 at F provided $\widehat{Y}_0$ is linear and

$$\text{MSPE}(\widehat{Y}_0, F) \leq \text{MSPE}(Y_0^\star, F) \tag{3.2.15}$$

for any other linear predictor $Y_0^\star$. Minimum MSPE linear predictors are sometimes called *best linear predictors* (BLPs).

Restricting further the class of predictors to those that are both linear and unbiased, one can again seek optimal MSPE predictors. To apply such a strategy, one must first determine which linear predictors are unbiased. Recall that unbiasedness is determined with respect to a family $\mathcal{F}$ of distributions. In the computer experiment literature, the emphasis is on finding a linear predictor $\widehat{Y}_0 = a_0 + \boldsymbol{a}^\top \boldsymbol{Y}^n$ that is unbiased with respect to every F in some family of distributions $\mathcal{F}$ and simultaneously minimizes the MSPE at F in the same family $\mathcal{F}$. Given $\mathcal{F}$, a predictor $\widehat{Y}_0 = a_0 + \boldsymbol{a}^\top \boldsymbol{Y}^n$ which is unbiased for $\mathcal{F}$ that satisfies (3.2.15) for $F \in \mathcal{F}$ is said to be *minimum MSPE linear unbiased* or simply a *best linear unbiased predictor* (BLUP).

Example 3.5 Consider best linear unbiased prediction for the nonparametric location parameter model (3.2.2) introduced in Example 3.1 where β_0 is *fixed*, i.e., for the family of distributions $\mathcal{F} = \mathcal{F}(\beta_0)$. Recall that $\widehat{Y}_0 = a_0 + \boldsymbol{a}^\top \boldsymbol{Y}^n$ is unbiased provided $a_0 + \beta_0 \sum_{i=1}^n a_i = \beta_0$. The MSPE of the unbiased predictor $\widehat{Y}_0 = a_0 + \boldsymbol{a}^\top \boldsymbol{Y}^n$ is

$$
\begin{aligned}
E\left\{\left(a_0 + \sum_{i=1}^n a_i Y_i - Y_0\right)^2\right\} &= E\left\{\left(a_0 + \sum_{i=1}^n a_i(\beta_0 + \epsilon_i) - \beta_0 - \epsilon_0\right)^2\right\} \\
&= \left(a_0 + \beta_0 \sum_{i=1}^n a_i - \beta_0\right)^2 \\
&\quad + \sigma_\epsilon^2 \times \sum_{i=1}^n a_i^2 + \sigma_\epsilon^2 \\
&= \sigma_\epsilon^2 \times \left(1 + \sum_{i=1}^n a_i^2\right) \qquad (3.2.16) \\
&\geq \sigma_\epsilon^2 \qquad (3.2.17)
\end{aligned}
$$

Equality holds in (3.2.16) because $\widehat{Y}_0$ is unbiased and equality occurs in (3.2.17) if and only if $a_0 = \beta_0$ and $a_1 = \ldots = a_n = 0$, which shows that

$$\widehat{Y}_0 = \beta_0$$

is the unique BLUP for model $\mathcal{F}$. For this example, as for previous examples that determined various types of best MSPE predictors, the BLUP depends heavily on $\mathcal{F}$.

Now consider the BLUP with respect to the enlarged model $\mathcal{F} = \mathcal{F}(\mathbb{R})$ where β_0 is an unknown real number and $\sigma_\epsilon^2 > 0$. For this $\mathcal{F}$, recall that every unbiased $\widehat{Y}_0 = a_0 + \boldsymbol{a}^\top \boldsymbol{Y}^n$ must satisfy $a_0 = 0$ and $\sum_{i=1}^n a_i = 1$. The MSPE of $\widehat{Y}_0$ is

$$
\begin{aligned}
E\left\{\left(\sum_{i=1}^n a_i Y_i - Y_0\right)^2\right\} &= \left(\beta_0 \sum_{i=1}^n a_i - \beta_0\right)^2 + \sigma_\epsilon^2 \times \sum_{i=1}^n a_i^2 + \sigma_\epsilon^2 \\
&= 0 + \sigma_\epsilon^2 \times \left(1 + \sum_{i=1}^n a_i^2\right) \qquad (3.2.18) \\
&\geq \sigma_\epsilon^2 (1 + 1/n), \qquad (3.2.19)
\end{aligned}
$$

where equality holds in (3.2.18) because $\sum_{i=1}^n a_i = 1$ and the minimum in (3.2.19) is calculated by observing that $\sum_{i=1}^n a_i^2$ is minimized subject to

$\sum_{i=1}^n a_i = 1$ when $a_i = 1/n$ for $1 \le i \le n$. This tells us that the sample mean $\widehat{Y}_0 = \frac{1}{n}\sum_{i=1}^n Y_i$ is the best linear unbiased predictor of Y_0 for the enlarged $\mathcal{F}$. The formula (3.2.19) for its MSPE is familiar from regression; σ_ϵ^2/n is the variance of the sample mean $\frac{1}{n}\sum_{i=1}^n Y_i$ while the "extra" σ_ϵ^2 accounts for the additional variability of Y_0. ■

Example 3.6 (BLUP for a measurement error model) Suppose that

$$Y_i \equiv Y(\boldsymbol{x}_i) = \sum_{j=1}^p f_j(\boldsymbol{x}_i)\beta_j + \epsilon_i = \boldsymbol{f}^\top(\boldsymbol{x}_i)\boldsymbol{\beta} + \epsilon_i$$

for $0 \le i \le n$, where the $\{f_j\}$ are known regression functions, $\boldsymbol{\beta} = (\beta_1, \ldots, \beta_p)^\top$ is unknown, and the measurement errors $\{\epsilon_i\}$ are uncorrelated with common mean zero and common variance σ_ϵ^2. Consider the BLUP of $Y_0 = Y(\boldsymbol{x}_0)$ for the moment model $\mathcal{F}$, where $\boldsymbol{\beta} \in \mathbb{R}^p$ and $\sigma_\epsilon^2 > 0$ but both are otherwise unknown. The predictor $\widehat{Y_0} = a_0 + \boldsymbol{a}^\top \boldsymbol{Y}^n$ is unbiased with respect to $\mathcal{F}$ provided

$$E\left\{a_0 + \boldsymbol{a}^\top \boldsymbol{Y}^n\right\} = a_0 + \boldsymbol{a}^\top \boldsymbol{F}\boldsymbol{\beta} \overset{\text{set}}{=} E\{Y_0\} = \boldsymbol{f}_0^\top \boldsymbol{\beta}$$

for all $(\boldsymbol{\beta}, \sigma_\epsilon^2)$, where $\boldsymbol{f}_0 = \boldsymbol{f}(\boldsymbol{x}_0)$. This is equivalent to

$$a_0 = 0 \quad \text{and} \quad \boldsymbol{F}^\top \boldsymbol{a} = \boldsymbol{f}_0. \tag{3.2.20}$$

In the Chapter Notes, Subsection 3.4, we show that the BLUP of Y_0 is

$$\widehat{Y_0} = \boldsymbol{f}_0^\top \widehat{\boldsymbol{\beta}}, \tag{3.2.21}$$

where $\widehat{\boldsymbol{\beta}} = (\boldsymbol{F}^\top \boldsymbol{F})^{-1}\boldsymbol{F}^\top \boldsymbol{Y}^n$ is the ordinary least squares estimator of $\boldsymbol{\beta}$ and that the BLUP is unique. ■

In the next section, we turn specifically to the problem of prediction for computer experiments. We begin our discussion with the Gaussian stochastic process model introduced in Section 2.3 and then derive predictors of $Y(\boldsymbol{x}_0)$ when $\boldsymbol{\beta}$ is unknown, first when the correlation function is *known* and then when it is *unknown*.

3.3 Prediction for Computer Experiments

Many types of analyses of computer experiment output are facilitated by having available easily-computed approximations to the computer code. Such approximations are often called *surrogates* in the global optimization literature (Booker, Dennis, Frank, Serafini, Torczon and Trosset (1999)) and *simulators* in the engineering literature (Bernardo et al. (1992)). Neural networks, splines, and predictors based on Gaussian process models are some of the approximation methods that have been used to form predictors for the output from computer experiments. We emphasize the latter for three reasons: the assumptions that lead to these predictors are explicitly stated; several familiar predictors, including linear and cubic splines, are special cases; and such predictors use data-dependent scaling of each input dimension.

The basis for most practical predictors is the Gaussian random function model introduced in Section 2.3. Recall that this model regards the deterministic computer output $y(\cdot)$ as the realization of the random function

$$Y(\boldsymbol{x}) = \sum_{j=1}^{p} f_j(\boldsymbol{x})\beta_j + Z(\boldsymbol{x}) = \boldsymbol{f}^\top(\boldsymbol{x})\boldsymbol{\beta} + Z(\boldsymbol{x}), \tag{3.3.1}$$

where $f_1(\cdot), \ldots, f_p(\cdot)$ are known regression functions, $\boldsymbol{\beta} = (\beta_1, \ldots, \beta_p)^\top$ is a vector of unknown regression coefficients, and $Z(\cdot)$ is a stationary Gaussian process on $\mathcal{X}$ having zero mean, variance σ_Z^2, and correlation function $R(\cdot)$.

Suppose that the training data consists of the computer output at the input sites $\boldsymbol{x}_1, \ldots, \boldsymbol{x}_n$ and that $y(\boldsymbol{x}_0)$ is to be predicted. The model (3.3.1) implies that $Y_0 = Y(\boldsymbol{x}_0)$ and $\boldsymbol{Y}^n = (Y(\boldsymbol{x}_1), \ldots, Y(\boldsymbol{x}_n))^\top$ has the multivariate normal distribution

$$\begin{pmatrix} Y_0 \\ \boldsymbol{Y}^n \end{pmatrix} \sim N_{1+n}\left[\begin{pmatrix} \boldsymbol{f}_0^\top \\ \boldsymbol{F} \end{pmatrix}\boldsymbol{\beta}, \sigma_Z^2 \begin{pmatrix} 1 & \boldsymbol{r}_0^\top \\ \boldsymbol{r}_0 & \boldsymbol{R} \end{pmatrix}\right], \tag{3.3.2}$$

where $\boldsymbol{f}_0 = \boldsymbol{f}(\boldsymbol{x}_0)$ is the $p \times 1$ vector of regression functions for $Y(\boldsymbol{x}_0)$, $\boldsymbol{F} = (f_j(\boldsymbol{x}_i))$ is the $n \times p$ matrix of regression functions for the training data, $\boldsymbol{r}_0 = (R(\boldsymbol{x}_0 - \boldsymbol{x}_1), \ldots, R(\boldsymbol{x}_0 - \boldsymbol{x}_n))^\top$ is the $n \times 1$ vector of correlations of $\boldsymbol{Y}^n$ with $Y(\boldsymbol{x}_0)$, and $\boldsymbol{R} = (R(\boldsymbol{x}_i - \boldsymbol{x}_j))$ is the $n \times n$ matrix of correlations among the $\boldsymbol{Y}^n$. The parameters $\boldsymbol{\beta} \in \mathbb{R}^p$ and $\sigma_Z^2 > 0$ are *unknown.*

In the following subsections we apply the methodology of Section 3.2 to find the BLUP of $Y(\boldsymbol{x}_0)$ under the following enlargement of the normal theory model (3.3.2). We *drop* the Gaussian assumption to make the model a nonparametric, moment model based on an arbitrary second-order stationary process; this model is denoted

$$\begin{pmatrix} Y_0 \\ \boldsymbol{Y}^n \end{pmatrix} \sim \left[\begin{pmatrix} \boldsymbol{f}_0^\top \\ \boldsymbol{F} \end{pmatrix}\boldsymbol{\beta}, \ \sigma_Z^2 \begin{pmatrix} 1 & \boldsymbol{r}_0^\top \\ \boldsymbol{r}_0 & \boldsymbol{R} \end{pmatrix}\right], \tag{3.3.3}$$

where $\boldsymbol{\beta}$ and $\sigma_Z^2 > 0$ are unknown. Subsection 3.3.1 treats the case of known correlation function $R(\cdot)$ and Subsection 3.3.2 considers the case of unknown $R(\cdot)$.

3.3.1 Prediction When the Correlation Function Is Known

If the correlation function $R(\cdot)$ is known, we show that the BLUP of $Y(\boldsymbol{x}_0)$ is

$$\widehat{Y}(\boldsymbol{x}_0) = \widehat{Y}_0 \equiv \boldsymbol{f}_0^\top \widehat{\boldsymbol{\beta}} + \boldsymbol{r}_0^\top \boldsymbol{R}^{-1}(\boldsymbol{Y}^n - \boldsymbol{F}\widehat{\boldsymbol{\beta}}), \tag{3.3.4}$$

where $\widehat{\boldsymbol{\beta}} = (\boldsymbol{F}^\top \boldsymbol{R}^{-1}\boldsymbol{F})^{-1}\boldsymbol{F}^\top \boldsymbol{R}^{-1}\boldsymbol{Y}^n$ is the generalized least squares estimator of $\boldsymbol{\beta}$. Of course, both $\boldsymbol{r}_0$ and $\boldsymbol{R}$ are specified when the correlation function $R(\cdot)$ is known.

In Section 3.2 we proved that (3.3.4) is the best MSPE predictor among all predictors under a two-stage model whose first stage was the Gaussian model (3.3.2). In this subsection, we *increase* the size of the model class but *restrict* the class of predictors to linear predictors that are unbiased with respect to the moment model (3.3.3).

Before showing that (3.3.4) is the BLUP under this model, we describe some of its properties under (3.3.3). First, $\widehat{Y}_0$ interpolates the training data $(\boldsymbol{x}_i, Y(\boldsymbol{x}_i))$ for $1 \leq i \leq n$ (Section 3.2).

Second, (3.3.4) is a LUP of $Y(\boldsymbol{x}_0)$; as shown below, this fact permits the straightforward derivation of its variance optimality. Linearity follows by substituting $\widehat{\boldsymbol{\beta}}$ into $\widehat{Y}_0$ yielding

$$\begin{aligned}\widehat{Y}_0 &= \Big[\boldsymbol{f}_0^\top (\boldsymbol{F}^\top \boldsymbol{R}^{-1}\boldsymbol{F})^{-1}\boldsymbol{F}^\top \boldsymbol{R}^{-1} \\ &\quad + \boldsymbol{r}_0^\top \boldsymbol{R}^{-1}(\boldsymbol{I}_n - \boldsymbol{F}(\boldsymbol{F}^\top \boldsymbol{R}^{-1}\boldsymbol{F})^{-1}\boldsymbol{F}^\top \boldsymbol{R}^{-1})\Big]\boldsymbol{Y}^n \\ &\equiv \boldsymbol{a}_*^\top \boldsymbol{Y}^n, \qquad (3.3.5)\end{aligned}$$

where (3.3.5) defines $\boldsymbol{a}_*$. Unbiasedness with respect to (3.3.3) holds because for any $\boldsymbol{\beta} \in \mathbb{R}^p$ and every $\sigma_Z^2 > 0$,

$$\begin{aligned}E\{\widehat{Y}_0\} &= \boldsymbol{a}_*^\top E\{\boldsymbol{Y}^n\} \\ &= \boldsymbol{a}_*^\top \boldsymbol{F}\boldsymbol{\beta} \\ &= [\boldsymbol{f}_0^\top \boldsymbol{I}_n + \boldsymbol{r}_0^\top \boldsymbol{R}^{-1}(\boldsymbol{F} - \boldsymbol{F}\boldsymbol{I}_n)]\boldsymbol{\beta} \qquad (3.3.6) \\ &= \boldsymbol{f}_0^\top \boldsymbol{\beta} \\ &= E\{Y(\boldsymbol{x}_0)\},\end{aligned}$$

where (3.3.6) holds by substituting for $\boldsymbol{a}_*$ in (3.3.5).

Third, we determine the behavior of $\widehat{Y}_0 = \widehat{Y}(\boldsymbol{x}_0)$ as a function of $\boldsymbol{x}_0$. This can be easily discerned because (3.3.4) depends on $\boldsymbol{x}_0$ only through the $n \times 1$ vector $\boldsymbol{r}_0 = \boldsymbol{r}(\boldsymbol{x}_0) = (R(\boldsymbol{x}_0 - \boldsymbol{x}_1), \ldots, R(\boldsymbol{x}_0 - \boldsymbol{x}_n))^\top$ and $\boldsymbol{f}(\boldsymbol{x}_0)$. Hence

$$\widehat{Y}(\boldsymbol{x}_0) = \sum_{j=1}^{p} \widehat{\beta}_j f_j(\boldsymbol{x}_0) + \sum_{i=1}^{n} d_i R(\boldsymbol{x}_0 - \boldsymbol{x}_i), \qquad (3.3.7)$$

where $\boldsymbol{d} = (d_1, \ldots, d_n)^\top = \boldsymbol{R}^{-1}(\boldsymbol{Y}^n - \boldsymbol{F}\widehat{\boldsymbol{\beta}})$. In the special case $Y(\boldsymbol{x}) = \beta_0 + Z(\boldsymbol{x})$, $\widehat{Y}(\boldsymbol{x}_0)$ depends on $\boldsymbol{x}_0$ only through $R(\boldsymbol{x}_0 - \boldsymbol{x}_i)$. The "smoothness" characteristics of $\widehat{Y}(\boldsymbol{x}_0)$ are inherited from those of $R(\cdot)$. For $\boldsymbol{x}_0$ "near" any $\boldsymbol{x}_i$ (more precisely, in the limit as $\boldsymbol{x}_0$ approaches $\boldsymbol{x}_i$), the behavior of $\widehat{Y}(\boldsymbol{x}_0)$ depends on that of $R(\cdot)$ at the origin.

In the Chapter Notes, Subsection 3.4, we show that (3.3.4) is the BLUP of $Y(\boldsymbol{x}_0)$ with respect to the family of distributions (3.3.3), where $\boldsymbol{\beta} \in \mathbb{R}^p$ and $\sigma_Z^2 > 0$ are unknown.

Example 3.7 As in Example 3.3, suppose that

$$f(x) = e^{-1.4x}\cos(7\pi x/2),$$

a dampened cosine curve over $0 \le x \le 1$, is the true output function. Figure 3.3 shows $f(x)$ as a solid line and the set of $n = 7$ points that we previously introduced as training data. Example 3.3 emphasized the role of the residuals in interpreting the BLUP (3.3.4) of $Y(x_0)$. We complete this example by re-examining the BLUP, this time as a function of x_0. Because the known correlation function for this example is

$$R(h) = e^{-136.1\times h^2},$$

we have

$$\widehat{Y}(x_0) = \sum_{i=1}^{7} d_i \exp\{-136.1(x_i - x_0)^2\}, \tag{3.3.8}$$

where $\{x_i\}_{i=1}^{7}$ are inputs for the training data and $(d_1, \ldots, d_7)$ are calculated from the expression following (3.3.7). Figure 3.3 shows that $\widehat{Y}(x_0)$ does interpolate the training data. Because each exponential component of (3.3.8) is infinitely differentiable in x_0, $\widehat{Y}(x_0)$ is also infinitely differentiable in x_0. ■

3.3.2 Prediction When the Correlation Function is Unknown

The basic strategy is to predict $y(\boldsymbol{x}_0)$ by

$$\widehat{Y}(\boldsymbol{x}_0) = \widehat{Y}_0 \equiv \boldsymbol{f}_0^\top \widehat{\boldsymbol{\beta}} + \widehat{\boldsymbol{r}}_0^\top \widehat{\boldsymbol{R}}^{-1}\left(\boldsymbol{Y}^n - \boldsymbol{F}\widehat{\boldsymbol{\beta}}\right), \tag{3.3.9}$$

where $\widehat{\boldsymbol{\beta}} = (\boldsymbol{F}^\top \widehat{\boldsymbol{R}}^{-1} \boldsymbol{F})^{-1} \boldsymbol{F}^\top \widehat{\boldsymbol{R}}^{-1} \boldsymbol{Y}^n$ and the estimates $\widehat{\boldsymbol{R}}$ and $\widehat{\boldsymbol{r}}_0$ are determined from an *estimator* of the correlation function $R(\cdot)$. Such predictors are called *empirical best linear unbiased predictors* (EBLUPs) of $Y(\boldsymbol{x}_0)$, despite the fact that they are typically no longer linear in the training data $\boldsymbol{Y}^n$ ($\widehat{\boldsymbol{R}} = \widehat{\boldsymbol{R}}(\boldsymbol{Y}^n)$ and $\widehat{\boldsymbol{r}}_0 = \widehat{\boldsymbol{r}}_0(\boldsymbol{Y}^n)$ are usually highly nonlinear in $\boldsymbol{Y}^n$) nor need they be unbiased for $Y(\boldsymbol{x}_0)$ (although see Kackar and Harville (1984)). Different EBLUPs correspond to different estimators of $R(\cdot)$.

Virtually all estimators of the correlation function that have appeared in the literature assume that $R(\cdot) = R(\cdot|\boldsymbol{\psi})$, where $\boldsymbol{\psi}$ is a finite vector of parameters. As an example, the exponential correlation function

$$R(\boldsymbol{h}|\boldsymbol{\psi}) = \exp\left\{-\sum_{j=1}^{d} |h_j/\theta_j|^{p_j}\right\}$$

has d scale parameters $\theta_1, \ldots, \theta_d$ and d power parameters $p_1, \ldots, p_d$ so that $\boldsymbol{\psi} = (\theta_1, \ldots, \theta_d, p_1, \ldots, p_d)$ contains $2 \times d$ components. In this case,

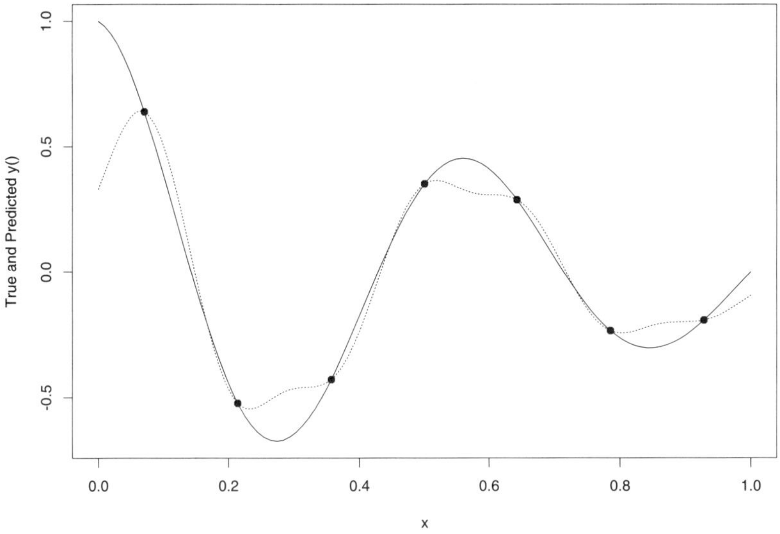

FIGURE 3.3. True curve $y(x) = \exp\{-1.4x\} \times \cos(3.5\pi x)$ (solid line), a seven point input design (dots), and the BLUP $\widehat{Y}(x_0)$ for $x_0 \in [0, 1.0]$ (dotted line)

the correlation matrix $\boldsymbol{R} = \boldsymbol{R}(\boldsymbol{\psi})$ depends on $\boldsymbol{\psi}$ as does the vector of correlations $\boldsymbol{r}_0 = \boldsymbol{r}_0(\boldsymbol{\psi})$. We describe four methods of estimating $\boldsymbol{\psi}$ that lead to four different EBLUPs. All except the "cross-validation" estimator of $\boldsymbol{\psi}$ assume that the training data have the Gaussian conditional distribution

$$[\boldsymbol{Y}^n | \boldsymbol{\beta}, \sigma_Z^2, \boldsymbol{\psi}] \sim N_n \left[\boldsymbol{F}\boldsymbol{\beta}, \sigma_Z^2 \boldsymbol{R}\right]. \tag{3.3.10}$$

Furthermore, the Bayes predictor assumes that prior information is available concerning model parameters.

We focus on the estimation of the correlation parameters $\boldsymbol{\psi}$ and not the process variance, σ_Z^2. This is because the predictor $\widehat{Y}_0$ depends only on $\boldsymbol{\psi}$ and is independent of σ_Z^2. However, in Section 4.1 (for example, formula (4.1.12)), we will see that σ_Z^2 is required to estimate the posterior variance of the predictor at each new training site $\boldsymbol{x}_0$. Except for cross validation, all the methods we present below can be used to estimate σ_Z^2.

Maximum Likelihood EBLUPs

Perhaps the most popular choice of $\boldsymbol{\psi}$ estimator is the maximum likelihood estimate (MLE). Using the multivariate normal assumption, the log likelihood is (up to an additive constant)

$$\ell(\boldsymbol{\beta}, \sigma_Z^2, \boldsymbol{\psi}) = -\frac{1}{2}\left[n \log \sigma_Z^2 + \log(\det(\boldsymbol{R})) + (\boldsymbol{y}^n - \boldsymbol{F}\boldsymbol{\beta})^\top \boldsymbol{R}^{-1}(\boldsymbol{y}^n - \boldsymbol{F}\boldsymbol{\beta})/\sigma_Z^2\right], \tag{3.3.11}$$

where $\det(\boldsymbol{R})$ denotes the determinant of $\boldsymbol{R}$. Given $\boldsymbol{\psi}$, the MLE of $\boldsymbol{\beta}$ is its generalized least squares estimate

$$\widehat{\boldsymbol{\beta}} = \widehat{\boldsymbol{\beta}}(\boldsymbol{\psi}) = \left(\boldsymbol{F}^\top \boldsymbol{R}^{-1} \boldsymbol{F}\right)^{-1} \boldsymbol{F}^\top \boldsymbol{R}^{-1} \boldsymbol{y}^n \tag{3.3.12}$$

and the MLE of σ_Z^2 is

$$\widehat{\sigma_Z^2} = \widehat{\sigma_Z^2}(\boldsymbol{\psi}) = \frac{1}{n}\left(\boldsymbol{y}^n - \boldsymbol{F}\widehat{\boldsymbol{\beta}}\right)^\top \boldsymbol{R}^{-1}\left(\boldsymbol{y}^n - \boldsymbol{F}\widehat{\boldsymbol{\beta}}\right). \tag{3.3.13}$$

Substituting these values into Equation (3.3.11), we obtain that the maximum of (3.3.11) over $\boldsymbol{\beta}$ and σ_Z^2 is

$$\ell(\widehat{\boldsymbol{\beta}}, \widehat{\sigma_Z^2}, \boldsymbol{\psi}) = -\frac{1}{2}\left[n \log \widehat{\sigma_Z^2}(\boldsymbol{\psi}) + \log(\det(\boldsymbol{R}(\boldsymbol{\psi}))) + n\right],$$

which depends on $\boldsymbol{\psi}$ alone. Thus the MLE chooses $\widehat{\boldsymbol{\psi}}$ to minimize

$$n \log \widehat{\sigma_Z^2}(\boldsymbol{\psi}) + \log\left(\det\left(\boldsymbol{R}(\boldsymbol{\psi})\right)\right), \tag{3.3.14}$$

where $\widehat{\sigma_Z^2}$ is defined by (3.3.13). The predictor corresponding to $\widehat{\boldsymbol{\psi}}$ is called an MLE-EBLUP of $Y(\boldsymbol{x}_0)$.

For the Gaussian stochastic process model, the SAS procedure PROC Mixed and program `GaSP` (Gaussian Stochastic Process, Welch, Buck, Sacks, Wynn, Mitchell and Morris (1992)) can calculate MLEs of the parameters for the product power exponential correlation function. The program `PErK` (Parametric EmpiRical Kriging, Williams (2001)) can calculate the MLEs of the parameters for both the product power exponential and product Matérn correlation functions.

Restricted Maximum Likelihood EBLUPs

Again assume that $R(\cdot)$ (and hence $\boldsymbol{R}$ and $\boldsymbol{r}_0$) depends on an unknown finite vector of parameters $\boldsymbol{\psi}$. Restricted (residual) maximum likelihood estimation (REML) of variance and covariance parameters was introduced by Patterson and Thompson (1971) as a method of determining less biased estimates of such parameters than maximum likelihood estimation (see also Patterson and Thompson (1974)). Some authors use the term *marginal maximum likelihood estimates* for the same concept.

The REML estimator of $\boldsymbol{\psi}$ maximizes the likelihood of a maximal set of linearly independent combinations of the $\boldsymbol{Y}^n$ where each linear combination is orthogonal to $\boldsymbol{F}\boldsymbol{\beta}$, the mean vector of $\boldsymbol{Y}^n$. Assuming that $\boldsymbol{F}$ is of full column rank p, this method corresponds to choosing an $(n-p) \times n$ matrix

$\boldsymbol{C}$ of full row rank $n-p$ that satisfies $\boldsymbol{CF} = \boldsymbol{0}$, and the REML estimator of $\boldsymbol{\psi}$ is the maximizer of the likelihood of the transformed "data"

$$\boldsymbol{W} \equiv \boldsymbol{C}\boldsymbol{Y}^n \sim N\left[\boldsymbol{CF\beta} = \boldsymbol{0}, \sigma_Z^2 \boldsymbol{C}\boldsymbol{R}(\boldsymbol{\psi})\boldsymbol{C}^\top\right]. \tag{3.3.15}$$

Notice that $\boldsymbol{W}$ contains p fewer "observations" than $\boldsymbol{Y}^n$ but $\boldsymbol{W}$ has the advantage that these data contain none of the unknown parameters $\boldsymbol{\beta}$.

As an example, consider the simplest linear model setting, that of independent and identically distributed $N(\beta_0, \sigma^2)$ observations $Y_1, \ldots, Y_n$. In this case, $p = 1$. The MLE of σ^2 based on the $Y_1, \ldots, Y_n$ is $\sum_{i=1}^n (Y_i - \overline{Y})^2/n$, which is a (downward) biased estimator of σ^2. One set of linear combinations having the orthogonality property $\boldsymbol{CF} = \boldsymbol{0}$ is obtained as follows. Let $\overline{Y}$ be the mean of $Y_1, \ldots, Y_n$. The linear combinations $W_1 = Y_1 - \overline{Y}$, $\ldots$, $W_{n-1} = Y_{n-1} - \overline{Y}$ each have mean zero and correspond to multiplying $\boldsymbol{Y}^n$ by an easily described $(n-1) \times n$ matrix $\boldsymbol{C}$ having full row rank $n-1$. Maximizing the likelihood based on $W_1, \ldots, W_{n-1}$ and expressing the result in terms of $Y_1, \ldots, Y_n$ gives

$$\sum_{i=1}^n (Y_i - \overline{Y})^2/(n-1).$$

The $n-1$ divisor in the error sum of squares produces an unbiased estimator of σ_Z^2.

Returning to the general case, it can be shown that the REML estimator of $\boldsymbol{\psi}$ is independent of the choice of linear combinations used to construct $\boldsymbol{W}^n$ subject to the number of columns of $\boldsymbol{C}$ being maximal in the sense of $\boldsymbol{C}$ having rank $n-p$ (Harville (1974), Harville (1977)). With some algebra it can be shown that the REML estimator of σ_Z^2 is

$$\widetilde{\sigma_Z^2} = \frac{n}{n-p}\widehat{\sigma_Z^2} = \frac{1}{n-p}\left(\boldsymbol{y}^n - \boldsymbol{F}\widehat{\boldsymbol{\beta}}\right)^\top \boldsymbol{R}^{-1}\left(\boldsymbol{y}^n - \boldsymbol{F}\widehat{\boldsymbol{\beta}}\right),$$

where $\widehat{\sigma_Z^2}$ is the MLE, formula (3.3.13), of σ_Z^2 and the REML estimator of $\boldsymbol{\psi}$ is the minimizer of

$$(n-p)\log \widetilde{\sigma_Z^2} + \log(\det(\boldsymbol{R}(\boldsymbol{\psi}))). \tag{3.3.16}$$

Cross-Validation EBLUPs

Cross-validation is a popular method for choosing model parameters in parametric model settings. Important early references describing cross-validation are Allen (1974), Stone (1974), and Stone (1977); Hastie et al. (2001) summarize recent applications.

We again assume that the correlation function is parametric with $R(\cdot) = R(\cdot|\boldsymbol{\psi})$ so that $\boldsymbol{R} = \boldsymbol{R}(\boldsymbol{\psi})$ and $\boldsymbol{r}_0 = \boldsymbol{r}_0(\boldsymbol{\psi})$. For $i = 1, \ldots, n$ let $\widehat{Y}_{-i}(\boldsymbol{\psi})$

denote the predictor (3.3.9) of $y(\boldsymbol{x}_i)$ when $\boldsymbol{\psi}$ is the true correlation parameter based on all the data *except* $(\boldsymbol{x}_i, y(\boldsymbol{x}_i))$. The cross-validated estimator of $\boldsymbol{\psi}$ minimizes the empirical mean squared prediction error

$$\text{XV-PE}(\boldsymbol{\psi}) = \sum_{i=1}^{n} (\widehat{Y}_{-i}(\boldsymbol{\psi}) - y(\boldsymbol{x}_i))^2. \tag{3.3.17}$$

More general forms of the cross-validation criterion have been proposed by Golub, Heath and Wahba (1979) and Wahba (1980).

Posterior Mode EBLUPs

The motivation and form of the posterior mode EBLUP is as follows. Recall that the minimum MSPE predictor of $Y(\boldsymbol{x}_0)$ is $E\{Y(\boldsymbol{x}_0)|\boldsymbol{Y}^n\}$ (Theorem 3.2.1). As described in Subsection 2.3.5, in fully Bayesian settings where a prior is available for $(\boldsymbol{\beta}, \sigma_Z^2, \boldsymbol{\psi})$, this conditional mean can be very difficult to compute. To explain why, suppose conditionally given $(\boldsymbol{\beta}, \sigma_Z^2, \boldsymbol{\psi})$ that $\boldsymbol{Y}^n$ is from a GRF and that a prior is available for $[\boldsymbol{\beta}, \sigma_Z^2, \boldsymbol{\psi}]$. The minimum MSPE predictor is by

$$E\left\{Y(\boldsymbol{x}_0)|\boldsymbol{Y}^n\right\} = E\left\{E\left\{Y(\boldsymbol{x}_0)|\boldsymbol{Y}^n, \boldsymbol{\psi}\right\}|\boldsymbol{Y}^n\right\}, \tag{3.3.18}$$

where the inner expectation on the right-hand side of (3.3.18) is regarded as a function of $\boldsymbol{\psi}$ and the outer expectation is with respect to the (marginal) posterior distribution $[\boldsymbol{\psi}|\boldsymbol{Y}^n]$.

The inner expectation (3.3.18) can be calculated by

$$E\left\{Y(\boldsymbol{x}_0)|\boldsymbol{Y}^n, \boldsymbol{\psi}\right\} = E\left\{(Y(\boldsymbol{x}_0), \boldsymbol{\beta}, \sigma_Z^2)|\boldsymbol{Y}^n, \boldsymbol{\psi}\right\},$$

which assumes that the conditional $[(\boldsymbol{\beta}, \sigma_Z^2)|\boldsymbol{Y}^n, \boldsymbol{\psi}]$ is available and the integration over $(\boldsymbol{\beta}, \sigma_Z^2)$ has been performed. Subsection 4.1.3 gives several examples of closed-form expressions for $E\left\{Y(\boldsymbol{x}_0)|\boldsymbol{Y}^n, \boldsymbol{\psi}\right\}$. Even where it can be evaluated in closed form, this integrand is a very complicated function of $\boldsymbol{\psi}$. For example, $E\left\{Y(\boldsymbol{x}_0)|\boldsymbol{Y}^n, \boldsymbol{\psi}\right\}$ involves the determinant of the correlation matrix as one of several terms in the examples of Subsection 4.1.3.

Even if $E\left\{Y(\boldsymbol{x}_0)|\boldsymbol{Y}^n, \boldsymbol{\psi}\right\}$ is known, the density of $[\boldsymbol{\psi}|\boldsymbol{Y}^n]$ generally cannot be expressed in closed form. One simple-minded but nevertheless attractive approximation to the right-hand side of (3.3.18) is

$$E\{Y(\boldsymbol{x}_0)|\boldsymbol{Y}^n, \widehat{\boldsymbol{\psi}}\}, \tag{3.3.19}$$

where $\widehat{\boldsymbol{\psi}}$ is the posterior mode of $[\boldsymbol{\psi}|\boldsymbol{Y}^n]$. The posterior mode of $\boldsymbol{\psi}$ is that $\widehat{\boldsymbol{\psi}}$ that maximizes

$$[\boldsymbol{\psi}|\boldsymbol{Y}^n] = \frac{[\boldsymbol{Y}^n|\boldsymbol{\psi}][\boldsymbol{\psi}]}{[\boldsymbol{Y}^n]} \propto [\boldsymbol{Y}^n|\boldsymbol{\psi}][\boldsymbol{\psi}].$$

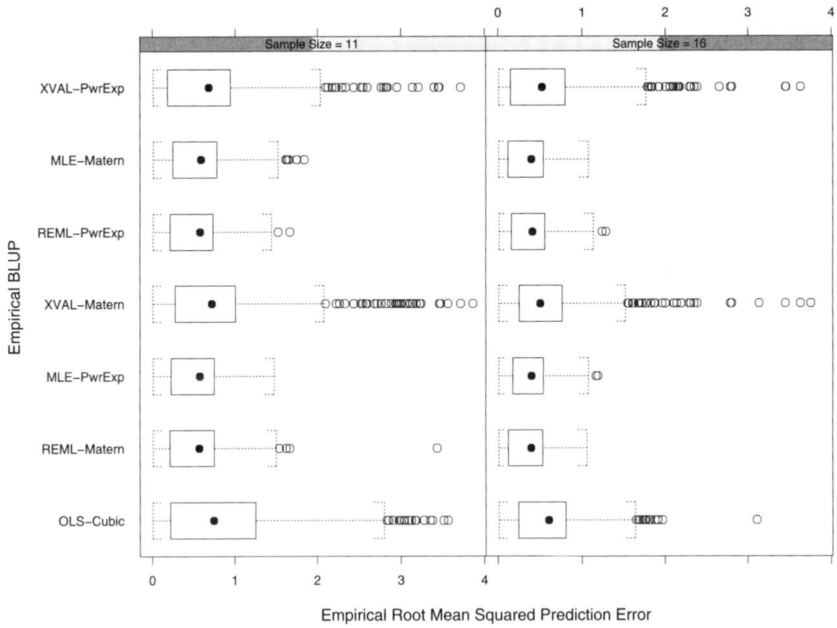

FIGURE 3.6. Distribution of ERMSPE over 625 equispaced grid of points in $[0,1]^2$ for the seven predictors conditioned on sample size (ignoring training data design and stochastic mechanism generating the true surface). Two large ERMSPE values by cross-validation EBLUPs were omitted from the plot.

outliers with slightly poorer performance occur more often with the Sobol´ design than with the maximin distance LHD.

Surface-by-surface comparisons of the ERMSPE of these predictors were made for the MLE-EBLUP versus the REML-EBLUP predictors and for the four different true surface groups. These plots showed little difference between the predictive ability of the two methods.

Before stating our primary conclusions, we wish to note several limitations of this particular empirical study that may effect our recommendations in some settings. First, all of the true surfaces are rather smooth, with the roughest corresponding to the krigifier with $\nu = 5$. Second, the dimension of the input in this study is $d = 2$, a rather low value. Third, none of the krigifier surfaces has a nonstationary trend term. It would be desirable to enhance the ranges of all three of these factors to broaden the applicability of our recommendation.

There are several additional caveats that should be kept in mind regarding our recommendations. This section makes recommendations based on the prediction accuracy of several predictors. Among the other important

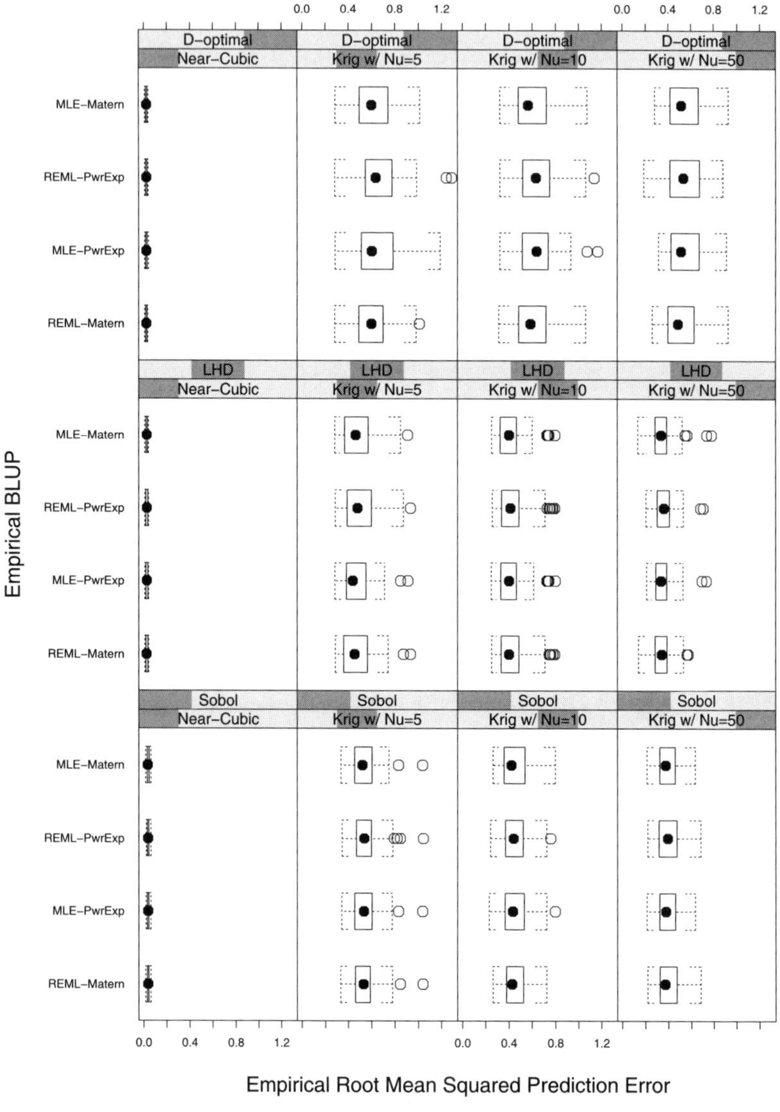

FIGURE 3.7. Distribution of ERMSPE over 625 equispaced grid of points in $[0,1]^2$ when $n = 16$ conditioned on training data design and type of true surface for the the MLE- and REML- EBLUPs based on either the power exponential and the Matérn correlation functions.

products of the prediction process are prediction bounds based on the plug-in estimates of prediction variability that will be introduced in Section 4.1.

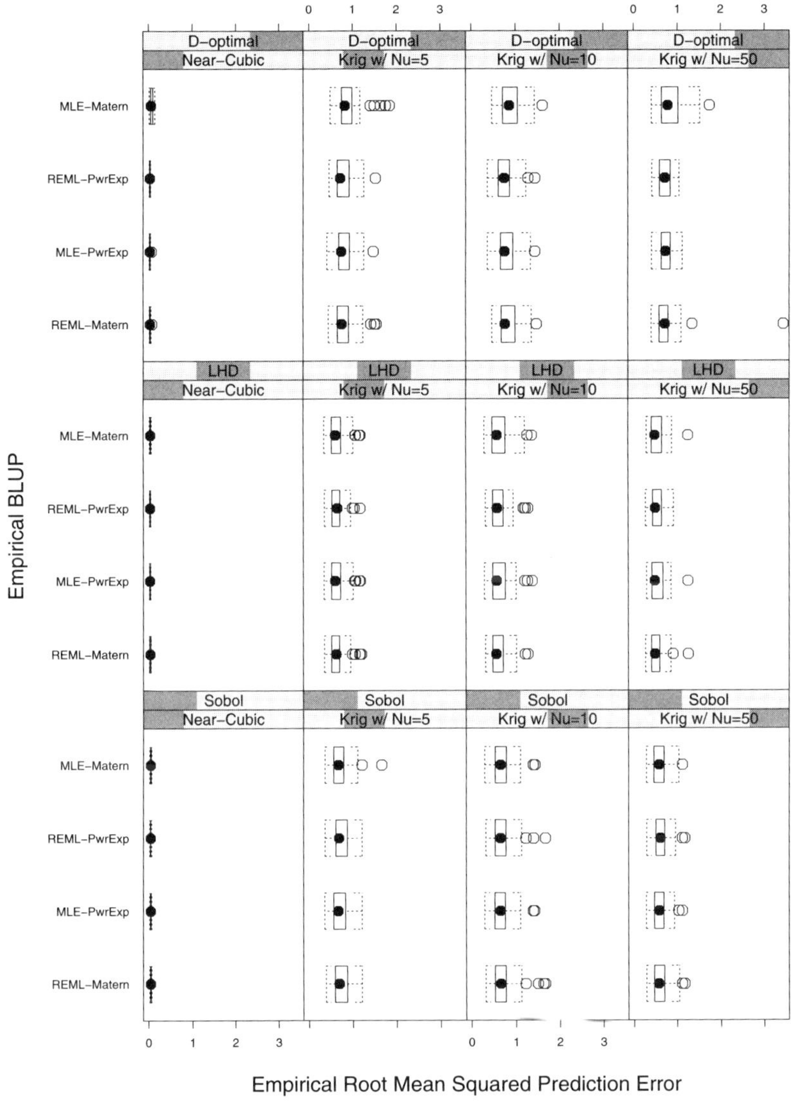

FIGURE 3.8. Distribution of ERMSPE over 625 equispaced grid of points in $[0, 1]^2$ when $n = 11$ conditioned on training data design and type of true surface for the MLE- and REML- EBLUPs based on either the power exponential and the Matérn correlation functions.

The accuracy of such prediction intervals will be addressed in more detail in Section 4.1. Our assessment of the empirical coverage of the corresponding intervals does not change the basic recommendations given below. Another issue is that we did not explicitly assess how small an initial sample size can be used to provide "reasonably" accurate prediction surfaces; for sequential designs (in addition to arising in high-dimensional, high-cost codes), such problems occur in Subsection 6.3.5 where the sequential design of a computer experiment is presented to find the global optimum of $y(\cdot)$. Certainly five observations per dimension appears to be adequate based on this limited study.

Recommendation *We recommend use of either the REML-EBLUP or MLE-EBLUP based on the power exponential correlation family. The Matérn correlation family produces similar ERMSPEs as the power exponential correlation family but is more computationally expensive. Maximin distance LHDs produce good predictors with Sobol´ designs a close second. D-optimal designs should be not be used to generate training data.*

Example 3.8 We illustrate the use of PErK to fit the REML empirical BLUP that is described earlier in this section. Recall the data introduced in Section 1.2 on the time for a fire to reach five feet above a fire source located in a room of a given *room height* and *room area.* In addition to room geometry, this output time is also dependent on the inputs: *heat loss fraction*, a measure of how well the room retains heat, and the *height of the fire source* above the room floor. Figure 3.9 plots each of the six two-dimensional projections of the 40 input points generated by a Sobol´ design. As noted in Chapter 5, Sobol´ designs provide points that have a greater range of inter-point distances than do the maximin distance Latin hypercube designs (see Example 5.8). This may allow better estimation of correlation parameters if the predictions are required at a set of points of varying distances from the training data.

Figure 3.10 displays scatterplots of each of the four input variables versus the time for a fire to reach five feet above the fire source, this output denoted by $y(\boldsymbol{x})$. Of these inputs, only *room area* appears to have a strong relationship with response time.

We desire to predict $y(\cdot)$ on a regular 320 point grid consisting of $4 \times 4 \times 4 \times 5$ equally spaced points over the ranges of the variables: heat loss fraction, room height, fire height, and room area, respectively. Our predictor is an EBLUP based on the Gaussian Stochastic Process with Matérn correlation function

$$R(\boldsymbol{h}) = \prod_{i=1}^{4} \frac{1}{\Gamma(\nu)2^{\nu-1}} \left(\frac{2\sqrt{\nu}\,|h_i|}{\theta_i} \right)^{\nu} K_{\nu} \left(\frac{2\sqrt{\nu}\,|h_i|}{\theta_i} \right) \qquad (3.3.25)$$

with unknown correlation parameter $\boldsymbol{\psi} = (\theta_1, \ldots, \theta_4, \nu)$. Recall that for large ν, the ith component correlation function of this product converges reml.corpar output file for Example 3.8

```
Correlation Family = Matern I

REML estimates of the correlation range parameters are:

Case Range
1          12.34622
2          6.88973
3          9.52963
4          7.08559

The REML of the correlation smoothness parameter is:
1.67681
```

The REML estimate of ν is 1.68 while the REMLs of the scale parameters $\theta_1, \ldots, \theta_4$ range from 6.89 to 12.35. ■

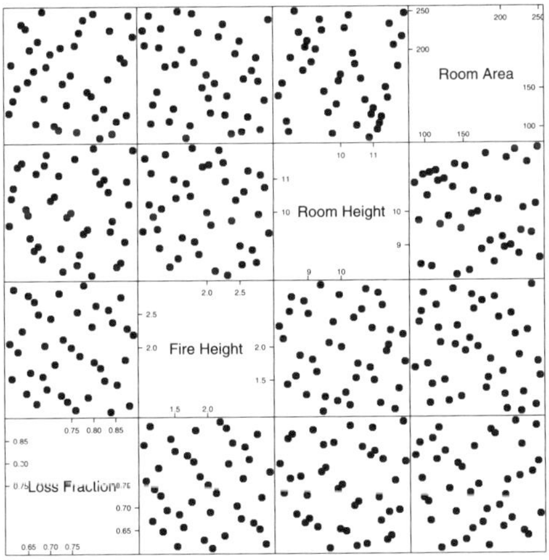

FIGURE 3.9. Scatterplot matrix of the 40 input points used in Example 3.8.

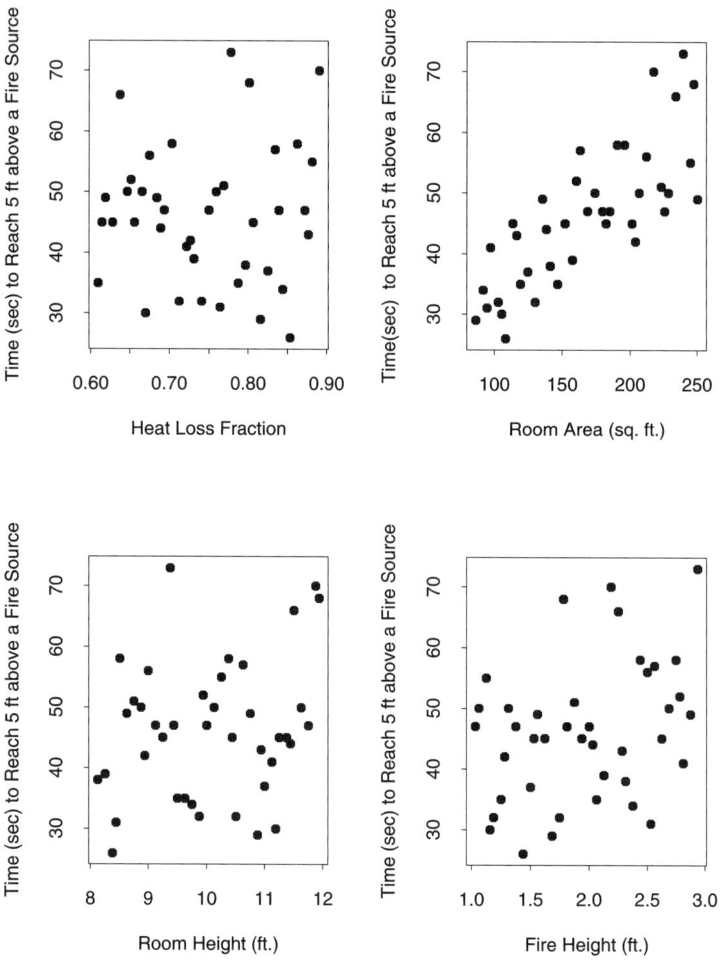

FIGURE 3.10. Scatterplots of the time for a fire to reach five feet above a fire source versus each of the inputs: (1) room height, (2) room area, (3) heat loss fraction, and (4) height of the fire source above the floor, using the data from Example 3.8.

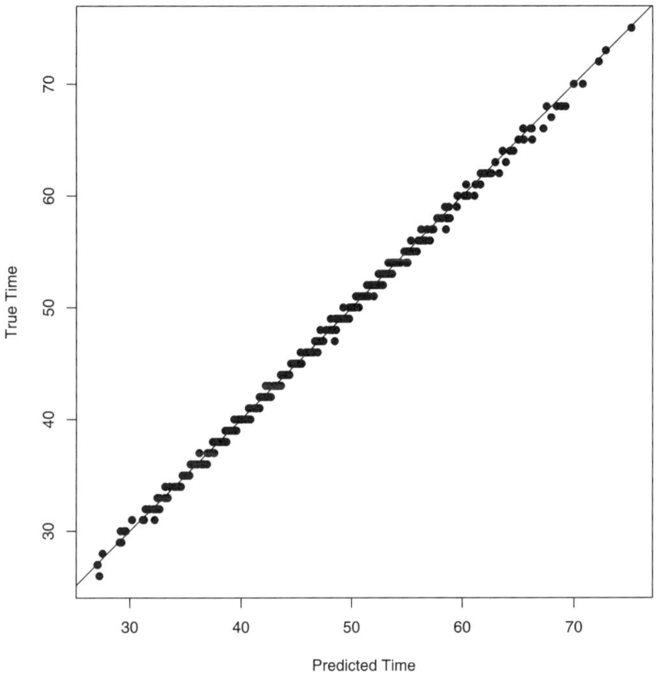

FIGURE 3.11. Scatterplot of the true versus predicted times to reach five feet above a fire source for the equispaced grid of 320 points used in Example 3.8.

3.4 Chapter Notes

3.4.1 Proof That (3.2.21) Is a BLUP (page 61)

The predictor $\widehat{Y_0}$ is linear because

$$\widehat{Y_0} = \boldsymbol{f}_0^\top (\boldsymbol{F}^\top \boldsymbol{F})^{-1} \boldsymbol{F}^\top \boldsymbol{Y}^n = \boldsymbol{a}_\star^\top \boldsymbol{Y}^n \, . \tag{3.4.1}$$

Furthermore $\widehat{Y_0}$ is unbiased because

$$\boldsymbol{F}^\top \boldsymbol{a}_\star = \boldsymbol{F}^\top \boldsymbol{F} (\boldsymbol{F}^\top \boldsymbol{F})^{-1} \boldsymbol{f}_0 = \boldsymbol{f}_0.$$

To see that (3.4.1) minimizes the MSPE pick any $\boldsymbol{a}$ for which $\boldsymbol{a}^\top \boldsymbol{Y}^n$ is unbiased, i.e., any $\boldsymbol{a}$ for which $\boldsymbol{F}^\top \boldsymbol{a} = \boldsymbol{f}_0$, and fix any $(\boldsymbol{\beta}, \sigma_\epsilon^2)$. Then the MSPE for this moment model is

$$\begin{aligned}
E\left\{\left(\boldsymbol{a}^\top \boldsymbol{Y}^n - Y_0\right)^2\right\} &= E\left\{\left(\boldsymbol{a}^\top (\boldsymbol{F}\boldsymbol{\beta} + \boldsymbol{\epsilon}^n) - \boldsymbol{f}_0^\top \boldsymbol{\beta} - \epsilon_0\right)^2\right\} \\
&= E\left\{\left(\boldsymbol{\beta}^\top (\boldsymbol{F}^\top \boldsymbol{a} - \boldsymbol{f}_0) + \sum_{i=1}^{n} a_i \epsilon_i - \epsilon_0\right)^2\right\} \\
&= E\left\{\left(\sum_{i=1}^{n} a_i \epsilon_i - \epsilon_0\right)^2\right\} & (3.4.2) \\
&= \sigma_\epsilon^2 \left(\sum_{i=1}^{n} a_i^2 + 1\right) = \sigma_\epsilon^2 \left(\boldsymbol{a}^\top \boldsymbol{a} + 1\right). & (3.4.3)
\end{aligned}$$

Equality holds in (3.4.2) because $\boldsymbol{a}$ satisfies the unbiasedness condition (3.2.20) and equality holds in (3.4.3) because the measurement errors are uncorrelated. This shows that the BLUP corresponds to that choice of $\boldsymbol{a}$ that minimizes $\boldsymbol{a}^\top \boldsymbol{a}$ subject to $\boldsymbol{F}^\top \boldsymbol{a} = \boldsymbol{f}_0$. But for any such $\boldsymbol{a}$,

$$\begin{aligned}
\boldsymbol{a}^\top \boldsymbol{a} &= (\boldsymbol{a} - \boldsymbol{a}_\star + \boldsymbol{a}_\star)^\top (\boldsymbol{a} - \boldsymbol{a}_\star + \boldsymbol{a}_\star) \\
&= (\boldsymbol{a} - \boldsymbol{a}_\star)^\top (\boldsymbol{a} - \boldsymbol{a}_\star) + \boldsymbol{a}_\star^\top \boldsymbol{a}_\star \\
&\quad + 2 (\boldsymbol{a} - \boldsymbol{a}_\star)^\top \boldsymbol{a}_\star \\
&= (\boldsymbol{a} - \boldsymbol{a}_\star)^\top (\boldsymbol{a} - \boldsymbol{a}_\star) + \boldsymbol{a}_\star^\top \boldsymbol{a}_\star & (3.4.4) \\
&\geq \boldsymbol{a}_\star^\top \boldsymbol{a}_\star, & (3.4.5)
\end{aligned}$$

where (3.4.1) defines $\boldsymbol{a}_\star$. Equality holds in (3.4.4) because the cross product is zero when $\boldsymbol{F}^\top \boldsymbol{a} = \boldsymbol{f}_0$:

$$\begin{aligned}
\boldsymbol{a} - \boldsymbol{a}_\star = \boldsymbol{a} - \boldsymbol{F}\left(\boldsymbol{F}^\top \boldsymbol{F}\right)^{-1} \boldsymbol{f}_0 &= \boldsymbol{a} - \boldsymbol{F}\left(\boldsymbol{F}^\top \boldsymbol{F}\right)^{-1} \boldsymbol{F}^\top \boldsymbol{a} \\
&= \left(\boldsymbol{I} - \boldsymbol{F}\left(\boldsymbol{F}^\top \boldsymbol{F}\right)^{-1} \boldsymbol{F}^\top\right) \boldsymbol{a}
\end{aligned}$$

which implies

$$\begin{aligned}(\boldsymbol{a}-\boldsymbol{a}_\star)^\top \boldsymbol{a}_\star &= \boldsymbol{a}^\top\left(\boldsymbol{I}-\boldsymbol{F}\left(\boldsymbol{F}^\top\boldsymbol{F}\right)^{-1}\boldsymbol{F}^\top\right)\times\left(\boldsymbol{F}\left(\boldsymbol{F}^\top\boldsymbol{F}\right)^{-1}\boldsymbol{f}_0\right)\\ &= \boldsymbol{a}^\top\left(\boldsymbol{a}_\star-\boldsymbol{F}\left(\boldsymbol{F}^\top\boldsymbol{F}\right)^{-1}\left(\boldsymbol{F}^\top\boldsymbol{F}\right)\left(\boldsymbol{F}^\top\boldsymbol{F}\right)^{-1}\boldsymbol{f}_0\right)\\ &= 0\,.\end{aligned}$$

Furthermore this argument shows that the BLUP is unique because equality holds in (3.4.5) if and only if $\boldsymbol{a}=\boldsymbol{a}_\star$. □

3.4.2 Proof That (3.3.4) Is a BLUP (page 63)

This proof is more complicated than its measurement error counterpart studied in Example 3.6 of Section 3.2. However, part of the argument used in Example 3.6 can be retained here. The class of LUPs of $Y(\boldsymbol{x}_0)$ with respect to (3.3.3) depends only on the first moment of $(Y_0, \boldsymbol{Y}^n)$ and hence is the same as for Example 3.6. The predictor $\widehat{Y}(\boldsymbol{x}_0) = a_0 + \boldsymbol{a}^\top \boldsymbol{Y}^n$ is unbiased for $Y(\boldsymbol{x}_0)$ provided

$$a_0 = 0 \quad \text{and} \quad \boldsymbol{F}^\top \boldsymbol{a} = \boldsymbol{f}_0. \tag{3.4.6}$$

Now fix any LUP of $Y(\boldsymbol{x}_0)$, say $\boldsymbol{a}^\top\boldsymbol{Y}^n$. Let $\boldsymbol{Z}^n = (Z(\boldsymbol{x}_1),\ldots,Z(\boldsymbol{x}_n))^\top$ and $Z_0 = Z(\boldsymbol{x}_0)$ be the corresponding stochastic process parts of $\boldsymbol{Y}^n$ and $Y(\boldsymbol{x}_0)$ in (3.3.1), respectively. For fixed $\boldsymbol{\beta}$ and σ_Z^2, the MSPE of $\boldsymbol{a}^\top\boldsymbol{Y}^n$ is

$$\begin{aligned}E\{(\boldsymbol{a}^\top\boldsymbol{Y}^n - Y_0)^2\} &= E\{(\boldsymbol{a}^\top(\boldsymbol{F}\boldsymbol{\beta}+\boldsymbol{Z}^n) - (\boldsymbol{f}_0^\top\boldsymbol{\beta}+Z_0))^2\}\\ &= E\{((\boldsymbol{a}^\top\boldsymbol{F}-\boldsymbol{f}_0^\top)\boldsymbol{\beta}+\boldsymbol{a}^\top\boldsymbol{Z}^n - Z_0)^2\}\\ &= E\{\boldsymbol{a}^\top\boldsymbol{Z}^n(\boldsymbol{Z}^n)^\top\boldsymbol{a}\\ &\quad -\ 2\boldsymbol{a}^\top\boldsymbol{Z}^n Z_0 + Z_0^2\} \qquad (3.4.7)\\ &= \sigma_Z^2\boldsymbol{a}^\top\boldsymbol{R}\boldsymbol{a} - 2\sigma_Z^2\boldsymbol{a}^\top\boldsymbol{r}_0 + \sigma_Z^2\\ &= \sigma_Z^2\left(\boldsymbol{a}^\top\boldsymbol{R}\boldsymbol{a} - 2\boldsymbol{a}^\top\boldsymbol{r}_0 + 1\right), \qquad (3.4.8)\end{aligned}$$

where (3.4.7) follows from (3.4.6). Thus the BLUP chooses $\boldsymbol{a}$ to minimize

$$\boldsymbol{a}^\top\boldsymbol{R}\boldsymbol{a} - 2\boldsymbol{a}^\top\boldsymbol{r}_0 \tag{3.4.9}$$

subject to

$$\boldsymbol{F}^\top\boldsymbol{a} = \boldsymbol{f}_0\,. \tag{3.4.10}$$

The method of Lagrange multipliers can be used to minimize the quadratic objective function (3.4.9) subject to linear constraints (3.4.10). We find $(\boldsymbol{a},\boldsymbol{\lambda}) \in \mathbb{R}^{n+p}$ to minimize

$$\boldsymbol{a}^\top\boldsymbol{R}\boldsymbol{a} - 2\boldsymbol{a}^\top\boldsymbol{r}_0 + 2\boldsymbol{\lambda}^\top(\boldsymbol{F}^\top\boldsymbol{a} - \boldsymbol{f}_0)\,. \tag{3.4.11}$$

Calculating the gradient of (3.4.11) with respect to $(\boldsymbol{a}, \boldsymbol{\lambda})$ and setting it equal to the zero vector gives the system of equations

$$\begin{aligned} \boldsymbol{F}^\top \boldsymbol{a} - \boldsymbol{f}_0 &= \boldsymbol{0} \\ \boldsymbol{R}\boldsymbol{a} - \boldsymbol{r}_0 + \boldsymbol{F}\boldsymbol{\lambda} &= \boldsymbol{0} \end{aligned}$$

or

$$\begin{pmatrix} 0 & \boldsymbol{F}^\top \\ \boldsymbol{F} & \boldsymbol{R} \end{pmatrix} \begin{pmatrix} \boldsymbol{\lambda} \\ \boldsymbol{a} \end{pmatrix} = \begin{pmatrix} \boldsymbol{f}_0 \\ \boldsymbol{r}_0 \end{pmatrix},$$

which implies

$$\begin{aligned} \begin{pmatrix} \boldsymbol{\lambda} \\ \boldsymbol{a} \end{pmatrix} &= \begin{pmatrix} 0 & \boldsymbol{F}^\top \\ \boldsymbol{F} & \boldsymbol{R} \end{pmatrix}^{-1} \begin{pmatrix} \boldsymbol{f}_0 \\ \boldsymbol{r}_0 \end{pmatrix} \\ &= \begin{pmatrix} -\boldsymbol{Q} & \boldsymbol{Q}\boldsymbol{F}^\top \boldsymbol{R}^{-1} \\ \boldsymbol{R}^{-1}\boldsymbol{F}\boldsymbol{Q} & \boldsymbol{R}^{-1} - \boldsymbol{R}^{-1}\boldsymbol{F}\boldsymbol{Q}\boldsymbol{F}^\top \boldsymbol{R}^{-1} \end{pmatrix} \times \begin{pmatrix} \boldsymbol{f}_0 \\ \boldsymbol{r}_0 \end{pmatrix}, \end{aligned}$$

where $\boldsymbol{Q} = (\boldsymbol{F}^\top \boldsymbol{R}^{-1} \boldsymbol{F})^{-1}$. After a small amount of algebra, the $\boldsymbol{a}$ solution gives (3.3.4) as the BLUP for the family (3.3.3). □

3.4.3 Implementation Issues

The calculation of either the MLE or the REML of the correlation parameters requires the repeated evaluation of the determinant and inverse of the $n \times n$ matrix $\boldsymbol{R}$. The Cholesky decomposition provides the most numerically stable method of calculating these quantities (Harville (1997)). Nevertheless, the repeated evaluation of these quantities is the most time consuming aspect of algorithms that sequentially add data. As an example, Williams, Santner and Notz (2000) report the times to maximize the REML likelihood which is required during the execution of their global optimization algorithm. In a six-dimensional input case, they fit the Matérn correlation function with a *single* shape parameter and separate range parameters for each input (a seven-dimensional $\boldsymbol{\psi}$ correlation parameter). When 50 training points were used, their optimization of the $\boldsymbol{\psi}$ likelihood (3.3.16) required 2,140 seconds of Sun Ultra 5 CPU time and this optimization required 4,230 seconds of CPU time for 82 training points. Fitting the power exponential model was faster with 1,105 seconds of CPU time required for the 50 point case and 3,100 seconds of CPU time for the 82 point case. Indeed, applications that require a sequence of correlation parameter estimates for increasing n often re-estimate these parameters only periodically, for example, when every fifth point is added to the design. A more rational plan is to re-estimate the correlation parameters more often for small n and more frequently for large n. For sufficiently large n, these estimators become intractable to calculate.

The dimension of the optimization problems required to find MLEs and REMLs can be large. For example, in a product exponential model with 20 input variables, each having unknown scale and power parameters, $\boldsymbol{\psi}$ is 40-dimensional. Such high-dimensional likelihood surfaces tend to have many local maxima, making global optimization difficult.

A variety of algorithms have been successfully used to determine MLEs and REMLs of correlation parameters. Among these are the Nelder-Mead simplex algorithm (Nelder and Mead (1965)), branch and bound algorithms (Jones et al. (1998)), and stochastic global optimization algorithms (Rinnooy Kan and Timmer (1984)). As noted above, the primary feature of a successful algorithm is that it must be capable of handling many local maxima in order to find a global maximum. There has been limited head-to-head comparison of the efficiency of these algorithms in finding optima.

As an example, to address high-dimensional MLE and REML parameter estimation problems, Welch et al. (1992) proposed using a dimensionality reduction scheme to perform a series of presumably simpler optimizations. The idea is to make tractable the high-dimensional $\boldsymbol{\psi}$ minimization in (3.3.14) or (3.3.16) by constraining the number of free parameters allowed in the minimization; only "important" input variables are allowed to possess their own unconstrained correlation parameters. This method is illustrated for the power exponential correlation family (2.3.14) with $\boldsymbol{\psi} = (\theta_1, \ldots, \theta_d)$ and $p_1 = \cdots = p_d = 2$.

First, each of the d input variables is scaled to have the same range. At each stage of the process, let $\mathcal{C}$ denote the indices of the variables having *common* values of the correlation parameters for that step and let $\mathcal{C}_{-j} = \mathcal{C} - \{j\}$. Notice that *S-0* is an initialization step in the following meta-algorithm, while *S-1* and *S-2* are induction steps.

S-0 Set $\mathcal{C} = \{1, 2, \ldots, d\}$, i.e., $\psi_1 = \cdots = \psi_d = \psi$. Maximize (3.3.14) or (3.3.16) as a function of ψ and denote the resulting log likelihood by ℓ_0.

S-1 For each $j \in \mathcal{C}$, maximize (3.3.14) or (3.3.16) under the constraint that variables ψ_h with $h \in \mathcal{C}_{-j}$ have a common value and ψ_j varies freely. Denote the result by ℓ_j.

S-2 Let j^M denote the variable producing the largest increase in $\ell_j - \ell_0$ for $j \in \mathcal{C}$.

S-3 If $\ell_{j^M} - \ell_0$ represents a "significant" increase in the log likelihood as judged by a stopping criterion, then set $\mathcal{C} = \mathcal{C}_{-j^M}$, $\ell_0 = \ell_{j^M}$, and fix ψ_{j^M} at its value estimated in *S-1*. Continue the next iteration at Step *S-1*. Otherwise, stop the algorithm and take the correlation parameter estimates produced by the previous iteration.

Of course, variations are possible in this scheme. For example, two-dimensional optimizations are used in every cycle of *S-1* because all ψ_{j^M}

estimated in previous cycles are fixed in subsequent ones. Instead, one could allow these values to vary freely along with the new ψ_j in the next *S-1* step. Thus the number of variables in the maximization would increase.

3.4.4 Alternate Predictors

This chapter has focused on the use of empirical best linear unbiased prediction, also known as empirical kriging prediction in the geostatistics literature. Empirical kriging methodology becomes numerically unstable when the size of the training sample, n, is large because the predictor (3.3.9) requires the inversion of an $n \times n$ matrix, which can be near-singular for certain correlation functions and choices of inputs $\boldsymbol{x}$. While numerous authors have written code to make empirical kriging more efficient (see An and Owen (1999) for some analysis of the computational burden), there is a point beyond which empirical kriging cannot be used. Hence several other approaches have been investigated in the literature that are computationally simpler than empirical kriging. We mention two of these methods.

One method of prediction that leads to computationally simpler predictors is to use the Gaussian random field model with a "Markov random field" model for the dependence structure. The special structure of the resulting correlation matrix allows for its analytic inversion and the usual empirical kriging predictor (3.3.9) has a simple form. See Cressie (1993), page 364, for a summary of the properties of MRF-based predictors and for additional references.

An and Owen (1999) described a predictor that they dubbed "quasi-regression." Their method exploits the use of an orthogonal basis function system to relate the inputs to the computer output. These methods are extremely computationally efficient and a wide variety of basis systems can be used.

where

$$\mu_{0|n} = \mu_{0|n}(\boldsymbol{x}_0) = \boldsymbol{f}_0^\top \boldsymbol{\mu}_{\beta|n} + \boldsymbol{r}_0^\top \boldsymbol{R}^{-1}\left(\boldsymbol{y}^n - \boldsymbol{F}\boldsymbol{\mu}_{\beta|n}\right), \tag{4.1.4}$$

for

$$\boldsymbol{\mu}_{\beta|n} = \left(\frac{\boldsymbol{F}^\top \boldsymbol{R}^{-1}\boldsymbol{F}}{\sigma_Z^2} + \frac{\boldsymbol{V}_0^{-1}}{\tau^2}\right)^{-1} \left(\frac{\boldsymbol{F}^\top \boldsymbol{R}^{-1}\boldsymbol{y}^n}{\sigma_Z^2} + \frac{\boldsymbol{V}_0^{-1}\boldsymbol{b}_0}{\tau^2}\right), \tag{4.1.5}$$

and

$$\sigma^2_{0|n} = \sigma^2_{0|n}(\boldsymbol{x}_0) = \sigma_Z^2\left\{1 - (\boldsymbol{f}_0^\top, \boldsymbol{r}_0^\top)\begin{bmatrix} -\frac{\sigma_Z^2}{\tau^2}\boldsymbol{V}_0^{-1} & \boldsymbol{F}^\top \\ \boldsymbol{F} & \boldsymbol{R}\end{bmatrix}^{-1}\begin{pmatrix}\boldsymbol{f}_0 \\ \boldsymbol{r}_0\end{pmatrix}\right\}. \tag{4.1.6}$$

(b) If

$$[\boldsymbol{\beta}] \sim 1 \tag{4.1.7}$$

on $\mathbb{R}^p$, then Y_0 has the predictive distribution

$$[Y_0 \mid \boldsymbol{Y}^n = \boldsymbol{y}^n] \sim N_1\left[\mu_{0|n}, \sigma^2_{0|n}\right], \tag{4.1.8}$$

where $\mu_{0|n} = \mu_{0|n}(\boldsymbol{x}_0)$ and $\sigma^2_{0|n} = \sigma^2_{0|n}(\boldsymbol{x}_0)$ are given by (4.1.4) and (4.1.6), respectively, with the substitution $\widehat{\boldsymbol{\beta}} = \left(\boldsymbol{F}^\top \boldsymbol{R}^{-1}\boldsymbol{F}\right)^{-1}\boldsymbol{F}^\top \boldsymbol{R}^{-1}\boldsymbol{y}^n$ for $\boldsymbol{\mu}_{\beta|n}$ in (4.1.4) and the $p \times p$ zero matrix for $-\frac{\sigma_Z^2}{\tau^2}\boldsymbol{V}_0^{-1}$ in (4.1.6).

There are several interesting features of (a) and (b) that aid in their interpretation. First, there is a "continuity" in the priors and posteriors as $\tau^2 \to \infty$. Concerning the priors, we had earlier observed that the non-informative prior (4.1.7) is the limit of the normal prior (4.1.2) as $\tau^2 \to \infty$. On the posterior side and paralleling this prior convergence, it can be calculated that the posterior mean $\mu_{0|n}(\boldsymbol{x}_0)$ and the posterior variance $\sigma^2_{0|n}(\boldsymbol{x}_0)$ in (4.1.8) for the non-informative prior are the limits, as $\tau^2 \to \infty$ of posterior mean and variance for the informative normal prior, which are (4.1.4) and (4.1.6), respectively. A second interesting feature is that while the prior (4.1.7) is intuitively non-informative, it is *not* a proper distribution. Nevertheless, the corresponding predictive distribution is proper. Indeed, we saw in (3.2.12) that $\widehat{\boldsymbol{\beta}}$ is the posterior mean of $\boldsymbol{\beta}$ given the data $\boldsymbol{Y}^n = \boldsymbol{y}^n$ for this two-stage model (and is the generalized least squares estimator of $\boldsymbol{\beta}$ from a frequentist viewpoint). Lastly, recall that in Subsection 3.3.2 we used a conditioning argument to derive the formula

$$\boldsymbol{f}_0^\top \widehat{\boldsymbol{\beta}} + \boldsymbol{r}_0^\top \boldsymbol{R}^{-1}\left(\boldsymbol{y}^n - \boldsymbol{F}\widehat{\boldsymbol{\beta}}\right)$$

in (4.1.8) as the predictive mean for the non-informative prior. This same type of conditioning can be applied to derive posterior mean (4.1.4)–(4.1.5) for the normal prior (4.1.2) for $\boldsymbol{\beta}$.

To understand the implications of Theorem 4.1.1, we examine some properties of the mean and variance of the predictive distribution (4.1.3). Both $\mu_{0|n}(\boldsymbol{x}_0)$ and $\sigma^2_{0|n}(\boldsymbol{x}_0)$ depend on $\boldsymbol{x}_0$ only through the regression functions $\boldsymbol{f}_0 = \boldsymbol{f}(\boldsymbol{x}_0)$ and the correlation vector $\boldsymbol{r}_0 = \boldsymbol{r}(\boldsymbol{x}_0)$. Focusing first on $\mu_{0|n}(\boldsymbol{x}_0)$, a little algebra shows that $\mu_{0|n}(\boldsymbol{x}_0)$ is linear in $\boldsymbol{Y}^n$ and, with additional calculation, that it is an unbiased predictor of $Y(\boldsymbol{x}_0)$, i.e., $\mu_{0|n}(\boldsymbol{x}_0)$ is a linear unbiased predictor of $Y(\boldsymbol{x}_0)$.

Second, the continuity and other smoothness properties of $\mu_{0|n}(\boldsymbol{x}_0)$ are inherited from those of the correlation function $R(\cdot)$ and the regressors $\{f_j(\cdot)\}_{j=1}^p$ because

$$\mu_{0|n}(\boldsymbol{x}_0) = \sum_{j=1}^{p} f_j(\boldsymbol{x}_0)\mu_{\boldsymbol{\beta}|n,j} + \sum_{i=1}^{n} d_i R(\boldsymbol{x}_0 - \boldsymbol{x}_i),$$

where $\mu_{\boldsymbol{\beta}|n,j}$ is the jth element of $\boldsymbol{\mu}_{\boldsymbol{\beta}|n}$. Previously, Subsection 3.3.1 had observed a parallel behavior for the BLUP (3.3.4), which is exactly the predictive mean $\mu_{0|n}(\boldsymbol{x}_0)$ in part (b) of Theorem 4.1.1.

Third, $\mu_{0|n}(\boldsymbol{x}_0)$ depends on the parameters σ^2_Z and τ^2 only through their ratio. This is because

$$\begin{aligned}
\boldsymbol{\mu}_{\boldsymbol{\beta}|n} &= \left(\frac{\boldsymbol{F}^\top \boldsymbol{R}^{-1}\boldsymbol{F}}{\sigma^2_Z} + \frac{\boldsymbol{V}_0^{-1}}{\tau^2}\right)^{-1}\left(\frac{\boldsymbol{F}\boldsymbol{R}^{-1}\boldsymbol{y}_n}{\sigma^2_Z} + \frac{\boldsymbol{V}_0^{-1}\boldsymbol{b}_0}{\tau^2}\right) \\
&= (\sigma^2_Z)\left(\boldsymbol{F}^\top \boldsymbol{R}^{-1}\boldsymbol{F} + \frac{\sigma^2_Z}{\tau^2}\boldsymbol{V}_0^{-1}\right)^{-1} \\
&\quad\times (\sigma^2_Z)^{-1}\left(\boldsymbol{F}^\top \boldsymbol{R}^{-1}\boldsymbol{y}^n + \frac{\sigma^2_Z}{\tau^2}\boldsymbol{V}_0^{-1}\boldsymbol{b}_0\right) \\
&= \left(\boldsymbol{F}^\top \boldsymbol{R}^{-1}\boldsymbol{F} + \frac{\sigma^2_Z}{\tau^2}\boldsymbol{V}_0^{-1}\right)^{-1}\left(\boldsymbol{F}^\top \boldsymbol{R}^{-1}\boldsymbol{y}^n + \frac{\sigma^2_Z}{\tau^2}\boldsymbol{V}_0^{-1}\boldsymbol{b}_0\right).
\end{aligned}$$

Lastly, the mean predictors $\mu_{0|n}(\boldsymbol{x}_0)$ in Theorem 4.1.1 interpolate the training data. This is true because when $\boldsymbol{x}_0 = \boldsymbol{x}_i$ for some $i \subset \{1, \ldots, n\}$, $\boldsymbol{f}_0 = \boldsymbol{f}(\boldsymbol{x}_i)$, and $\boldsymbol{r}_0^\top \boldsymbol{R}^{-1} = \boldsymbol{e}_i^\top$, the ith unit vector. Thus

$$\begin{aligned}
\mu_{0|n}(\boldsymbol{x}_i) &= \boldsymbol{f}^\top(\boldsymbol{x}_i)\boldsymbol{\mu}_{\boldsymbol{\beta}|n} + \boldsymbol{r}_0^\top \boldsymbol{R}^{-1}(\boldsymbol{Y}^n - \boldsymbol{F}\boldsymbol{\mu}_{\boldsymbol{\beta}|n}) \\
&= \boldsymbol{f}^\top(\boldsymbol{x}_i)\boldsymbol{\mu}_{\boldsymbol{\beta}|n} + \boldsymbol{e}_i^\top(\boldsymbol{Y}^n - \boldsymbol{F}\boldsymbol{\mu}_{\boldsymbol{\beta}|n}) \\
&= \boldsymbol{f}^\top(\boldsymbol{x}_i)\boldsymbol{\mu}_{\boldsymbol{\beta}|n} + (Y_i - \boldsymbol{f}^\top(\boldsymbol{x}_i)\boldsymbol{\mu}_{\boldsymbol{\beta}|n}) \\
&= Y_i\,.
\end{aligned}$$

Example 4.1 This example illustrates the effect of various choices of the prior $[\boldsymbol{\beta}]$ on the mean of the predictive distribution which is stated in Theorem 4.1.1. We use the same true function

$$y(x) = e^{-1.4x}\cos(7\pi x/2)\,, \qquad 0 < x < 1,$$

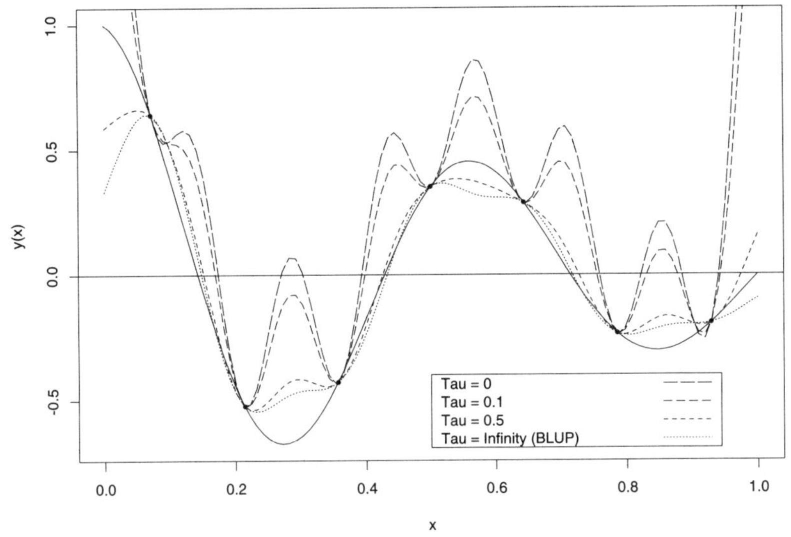

FIGURE 4.1. The predictor $\mu_{0|n} = \mu_{\beta_0|n} + \boldsymbol{r}_0^\top \boldsymbol{R}^{-1}(\boldsymbol{y}^n - \boldsymbol{1}_n \mu_{\beta_0|n})$ in (4.1.9) and (4.1.10) with $b_0 = 5$, $\sigma_Z = .41$, and four choices of τ^2.

and $n = 7$ point training data as in Examples 3.3 and 3.7. The predictive distribution of $Y(\boldsymbol{x}_0)$ is based on the two-stage model whose first stage is the stationary stochastic process

$$Y(x) \mid \beta_0 = \beta_0 + Z(x), \qquad 0 < x < 1,$$

where $\beta_0 \in \mathbb{R}$ and $R(h) = \exp\{-136.1 \times h^2\}$.

Suppose we take $\beta_0 \sim N(b_0, \tau^2 \times v_0^2)$ in part (a) of Theorem 4.1.1 and $v_0 = 1$ to guarantee identifiability of the prior variance. Both b_0 and τ^2 are assumed known. The mean of the posterior, Equation (4.1.4), is

$$\mu_{0|n} = \mu_{\beta_0|n} + \boldsymbol{r}_0^\top \boldsymbol{R}^{-1} \left(\boldsymbol{y}^n - \boldsymbol{1}_n \mu_{\beta_0|n}\right), \tag{4.1.9}$$

where $\mu_{\beta_0|n}$ is the posterior mean of β_0 given $\boldsymbol{Y}^n$ which is

$$\begin{aligned} \mu_{\beta_0|n} = \mu_{\beta_0|n}(b_0, \tau^2) &= \frac{(\boldsymbol{1}_n^\top \boldsymbol{R}^{-1} \boldsymbol{y}^n + b_0 \sigma_Z^2/\tau^2)}{(\boldsymbol{1}_n^\top \boldsymbol{R}^{-1} \boldsymbol{1}_n + \sigma_Z^2/\tau^2)} \\ &= \omega b_0 + (1-\omega)(\boldsymbol{1}_n^\top \boldsymbol{R}^{-1} \boldsymbol{1}_n^\top)^{-1} (\boldsymbol{1}_n^\top \boldsymbol{R}^{-1} \boldsymbol{y}^n) \\ &= \omega b_0 + (1-\omega)\widehat{\beta_0}, \end{aligned} \tag{4.1.10}$$

where $\omega = \sigma_Z^2/[\tau^2 \boldsymbol{1}_n^\top \boldsymbol{R}^{-1} \boldsymbol{1}_n + \sigma_Z^2] \in (0,1)$. In words, (4.1.10) can be interpreted as saying that the posterior mean of β_0 given $\boldsymbol{Y}^n$ is a convex combination of the MLE of β_0 and its prior mean, which are $\widehat{\beta_0}$, the generalized least squares estimator of β_0 and b_0, respectively. The behavior of the weight ω provides additional intuition about two extreme cases of $\mu_{\beta_0|n}$. When the prior certainty in b_0 increases in such a way that $\tau^2 \to 0$ for fixed process variance σ_Z^2, then $\omega \to 1$ and $\mu_{0|n} \to b_0$, meaning that the predictor uses only the prior and ignores the data, which is reasonable for perfect prior information. Similarly, when the prior certainty in b_0 decreases in such a way that $\tau^2 \to \infty$ for fixed process variance σ_Z^2, then $\omega \to 0$ and $\mu_{0|n} \to \widehat{\beta_0}$ so the predictor uses only the data and ignores the prior, which is, again, intuitively reasonable when there is no prior information.

Figure 4.1 shows the effect of changing the prior on $\mu_{0|n}(x_0)$; remember that $\mu_{0|n}(x_0)$ depends not only on $\mu_{\beta_0|n}$ but also on the correction term $\boldsymbol{r}_0^\top \boldsymbol{R}^{-1}(\boldsymbol{y}^n - \boldsymbol{1}_n\mu_{\beta_0|n})$. The four predictors correspond to $b_0 = 5$, $\sigma_Z = .41$, and four τ^2 values, with a fixed power exponential correlation function. *Smaller* τ^2 values produce predictors that have greater excursions from the data than do predictors having *greater* τ^2 values. In this case, the predictors having smaller τ^2 produce larger excursions from the true curve than does the BLUP (3.3.4) (which equals $\mu_{0|n}(x_0)$ with $\tau^2 = \infty$). This prior mean of β_0 was purposely taken to be the "large" value of $b_0 = 5.0$ which is not near the data to illustrate the effect of τ^2. Smaller τ^2 values correspond to being more certain about the prior and thus, the predictor pulls away from the data except when the training data pull it back. ■

Turning attention to the variance of the predictive distribution, $\sigma^2_{0|n}(\boldsymbol{x}_0)$, first observe that this quantity can be interpreted as the (unconditional) mean squared prediction error of $\mu_{0|n}(\boldsymbol{x}_0)$ because

$$\begin{aligned}
\text{MSPE}(\mu_{0|n}(\boldsymbol{x}_0)) &= E\left\{\left(Y(\boldsymbol{x}_0) - \mu_{0|n}(\boldsymbol{x}_0)\right)^2\right\} \\
&= E\left\{(Y(\boldsymbol{x}_0) - E\{Y(\boldsymbol{x}_0)|\boldsymbol{Y}^n\})^2\right\} \\
&= E\left\{E\left\{(Y(\boldsymbol{x}_0) - E\{Y(\boldsymbol{x}_0)|\boldsymbol{Y}^n\})^2 \,|\boldsymbol{Y}^n\right\}\right\} \\
&= E\left\{\sigma^2_{0|n}(\boldsymbol{x}_0)\right\} \\
&= \sigma^2_{0|n}(\boldsymbol{x}_0).
\end{aligned}$$

Thus $\sigma^2_{0|n}(\boldsymbol{x}_0)$ is the usual measure of precision of $\mu_{0|n}(\boldsymbol{x}_0)$.

The reader should be alert to the fact that $\sigma^2_{0|n}(\boldsymbol{x}_0)$ has a number of equivalent algebraic forms that are used in different papers and books (see Sacks, Welch, Mitchell and Wynn (1989), Cressie (1993)). Using basic matrix manipulations and starting with (4.1.6), we obtain

$$\begin{aligned}
\sigma^2_{0|n} &= \sigma^2_Z \left\{ 1 - (\boldsymbol{f}_0^\top, \boldsymbol{r}_0^\top) \begin{bmatrix} -\frac{\sigma^2_Z}{\tau^2}\boldsymbol{V}_0^{-1} & \boldsymbol{F}^\top \\ \boldsymbol{F} & \boldsymbol{R} \end{bmatrix}^{-1} \begin{pmatrix} \boldsymbol{f}_0 \\ \boldsymbol{r}_0 \end{pmatrix} \right\} \\
&= \sigma^2_Z \left\{ 1 - \left[-\boldsymbol{f}_0^\top \boldsymbol{Q}^{-1} \boldsymbol{f}_0 + 2\boldsymbol{f}_0^\top \boldsymbol{Q}^{-1} \boldsymbol{F}^\top \boldsymbol{R}^{-1} \boldsymbol{r}_0 \right.\right. \\
&\qquad \left.\left. + \ \boldsymbol{r}_0^\top \{ \boldsymbol{R}^{-1} - \boldsymbol{R}^{-1} \boldsymbol{F} \boldsymbol{Q}^{-1} \boldsymbol{F}^\top \boldsymbol{R}^{-1} \} \boldsymbol{r}_0 \right] \right\} \qquad (4.1.11) \\
&= \sigma^2_Z \{ 1 - \boldsymbol{r}_0^\top \boldsymbol{R}^{-1} \boldsymbol{r}_0 + \boldsymbol{f}_0^\top \boldsymbol{Q}^{-1} \boldsymbol{f}_0 - 2\boldsymbol{f}_0^\top \boldsymbol{Q}^{-1} \boldsymbol{F}^\top \boldsymbol{R}^{-1} \boldsymbol{r}_0 \\
&\qquad + \boldsymbol{r}_0^\top \boldsymbol{R}^{-1} \boldsymbol{F} \boldsymbol{Q}^{-1} \boldsymbol{F}^\top \boldsymbol{R}^{-1} \boldsymbol{r}_0 \} \\
&= \sigma^2_Z \{ 1 - \boldsymbol{r}_0^\top \boldsymbol{R}^{-1} \boldsymbol{r}_0 + \boldsymbol{h}^\top \boldsymbol{Q}^{-1} \boldsymbol{h} \}, \qquad (4.1.12)
\end{aligned}$$

where

$$\boldsymbol{Q} = \boldsymbol{F}^\top \boldsymbol{R}^{-1} \boldsymbol{F} + \frac{\sigma^2_Z}{\tau^2} \boldsymbol{V}_0^{-1}, \qquad (4.1.13)$$

$\boldsymbol{h} = \boldsymbol{f}_0 - \boldsymbol{F}^\top \boldsymbol{R}^{-1} \boldsymbol{r}_0$, and (4.1.11) follows from Lemma B.3.1. In particular, expression (4.1.12)

$$\sigma^2_{0|n} = \sigma^2_Z \{ 1 - \boldsymbol{r}_0^\top \boldsymbol{R}^{-1} \boldsymbol{r}_0 + \boldsymbol{h}^\top (\boldsymbol{F}^\top \boldsymbol{R}^{-1} \boldsymbol{F})^{-1} \boldsymbol{h} \}$$

is a frequently-used expression for the variance of the BLUP (3.3.4), i.e., for $\mu_{0|n}(\boldsymbol{x}_0)$ in Part (b) of Theorem 4.1.1. (See, for example, (5.3.15) of Cressie (1993).)

Intuitively, the variance of the posterior of $Y(\boldsymbol{x}_0)$ given $Y(\boldsymbol{x}_1), \ldots, Y(\boldsymbol{x}_n)$ should be *zero* whenever $\boldsymbol{x}_0 = \boldsymbol{x}_i$ because we *know* exactly the response at each of the training data sites $\boldsymbol{x}_i$ and there is no measurement error term in our stochastic process model. To see that this *is* the case analytically, fix $\boldsymbol{x}_0 = \boldsymbol{x}_i$ for some $1 \le i \le n$, recall that $\boldsymbol{r}_0^\top \boldsymbol{R}^{-1} = \boldsymbol{e}_i^\top$, and observe that $\boldsymbol{f}_0 = \boldsymbol{f}(\boldsymbol{x}_i)$. From (4.1.12),

$$\begin{aligned}
\sigma^2_{0|n}(\boldsymbol{x}_i) &= \sigma^2_Z \{ 1 - \boldsymbol{r}_0^\top \boldsymbol{R}^{-1} \boldsymbol{r}_0 + (\boldsymbol{f}_0^\top - \boldsymbol{r}_0^\top \boldsymbol{R}^{-1} \boldsymbol{F}) \boldsymbol{Q}^{-1} (\boldsymbol{f}_0 - \boldsymbol{F}^\top \boldsymbol{R}^{-1} \boldsymbol{r}_0) \} \\
&= \sigma^2_Z \{ 1 - \boldsymbol{e}_i^\top \boldsymbol{r}(\boldsymbol{x}_i) + (\boldsymbol{f}^\top(\boldsymbol{x}_i) - \boldsymbol{e}_i^\top \boldsymbol{F}) \boldsymbol{Q}^{-1} (\boldsymbol{f}(\boldsymbol{x}_i) - \boldsymbol{F}^\top \boldsymbol{e}_i) \} \\
&= \sigma^2_Z \{ 1 - 1 + (\boldsymbol{f}^\top(\boldsymbol{x}_i) - \boldsymbol{f}^\top(\boldsymbol{x}_i)) \boldsymbol{Q}^{-1} (\boldsymbol{f}(\boldsymbol{x}_i) - \boldsymbol{f}(\boldsymbol{x}_i)) \} \\
&= \sigma_Z \{ 1 - 1 + 0 \} = 0
\end{aligned}$$

where $\boldsymbol{Q}$ is given in (4.1.13).

Perhaps the most important use of Theorem 4.1.1 is to provide pointwise predictive bands about the predictor $\mu_{0|n}(\boldsymbol{x}_0)$. The bands can be obtained by using the fact that

$$\frac{Y(\boldsymbol{x}_0) - \mu_{0|n}(\boldsymbol{x}_0)}{\sigma^2_{0|n}(\boldsymbol{x}_0)} \sim N(0, 1)\,.$$

This gives the posterior prediction interval

$$P\{Y(\boldsymbol{x}_0) \in \mu_{0|n}(\boldsymbol{x}_0) \pm \sigma_{0|n}(\boldsymbol{x}_0) z^{\alpha/2} | \boldsymbol{Y}^n\} = 1 - \alpha,$$

where $z^{\alpha/2}$ is the upper $\alpha/2$ critical point of the standard normal distribution (see Appendix A). As a special case, if the input x_0 is real with limits $a < x_0 < b$, then $\mu_{0|n}(x_0) \pm \sigma_{0|n}(x_0) z^{\alpha/2}$ are pointwise $100(1-\alpha)\%$ prediction bands for $Y(x_0)$, $a < x_0 < b$. Below, we illustrate the prediction band calculation following the statement of the predictive distribution for our second hierarchical $(Y_0, \boldsymbol{Y}^n)$ model in Theorem 4.1.2.

4.1.3 Predictive Distributions When $\boldsymbol{R}$ and $\boldsymbol{r}_0$ Are Known

Using the fact that $[\boldsymbol{\beta}, \sigma_Z^2] = [\boldsymbol{\beta} \,|\, \sigma_Z^2] \times [\sigma_Z^2]$, Theorem 4.1.2 provides the (predictive) distribution of $Y(\boldsymbol{x}_0)$ given $\boldsymbol{Y}^n$ for four priors corresponding to informative and non-informative choices for each of the terms $[\boldsymbol{\beta} \,|\, \sigma_Z^2]$ and $[\sigma_Z^2]$, i.e., proper and improper distributions, respectively. These four combinations give rise to the simplest $[\boldsymbol{\beta}, \sigma_Z^2]$ priors that are, with adjustments given in Subsection 4.1.4, useful in practical situations. In all cases, the posterior is a location shifted and scaled univariate t distribution having degrees of freedom that are enhanced when there is informative prior information for either $\boldsymbol{\beta}$ or σ_Z^2 (see Appendix B.2 for a definition of the non-central t distribution, $T_1(\nu, \mu, \sigma)$).

The informative conditional $[\boldsymbol{\beta} \,|\, \sigma_Z^2]$ choice is the multivariate normal distribution with known mean $\boldsymbol{b}_0$ and known correlation matrix $\boldsymbol{V}_0$; lacking more definitive information, $\boldsymbol{V}_0$ is often taken to be diagonal, if not simply the identity matrix. This model makes strong assumptions, for example, it says that, componentwise, $\boldsymbol{\beta}$ is equally likely to be less than or greater than $\boldsymbol{b}_0$. The non-informative $\boldsymbol{\beta}$ prior is the intuitive choice

$$\pi(\boldsymbol{\beta}) = 1.$$

Our informative prior for σ_Z^2 is the distribution of a constant divided by a Chi-square random variable, i.e., we model $[\sigma_Z^2]$ as having the density of the $c_0/\chi^2_{\nu_0}$ random variable. This density has prior mean and variance

$$\frac{c_0}{\nu_0 - 2}, \text{ for } \nu_0 > 2 \text{ and } \frac{2 \times c_0^2}{(\nu_0 - 2)^2(\nu_0 - 4)}, \text{ for } \nu_0 > 4,$$

which allows one to more easily assign the model parameters. The non-informative prior used below is "Jeffreys prior"

$$\pi(\sigma_Z^2) = \frac{1}{\sigma_Z^2}$$

(see Jeffreys (1961), who gives arguments for this choice). Table 4.1 lists the notation for each of these four combinations that is used in Theorem 4.1.2.

	$[\sigma_Z^2]$	
$[\boldsymbol{\beta} \mid \sigma_Z^2]$	$c_0/\chi^2_{\nu_0}$	$1/\sigma_Z^2$
$N(\boldsymbol{b}_0, \sigma_Z^2 \boldsymbol{V}_0)$	(1)	(2)
1	(3)	(4)

TABLE 4.1. Labels of four $[\boldsymbol{\beta}, \sigma_Z^2]$ priors corresponding to informative and non-informative choices for each of $[\boldsymbol{\beta} \mid \sigma_Z^2]$ and $[\sigma_Z^2]$.

Theorem 4.1.2 Suppose $(Y_0, \boldsymbol{Y}^n)$ follows a two-stage model in which the conditional distribution $[(Y_0, \boldsymbol{Y}^n) \mid (\boldsymbol{\beta}, \sigma_Z^2)]$ is given by (4.1.1) and $[(\boldsymbol{\beta}, \sigma_Z^2)]$ has one of the priors corresponding to the four products (1)–(4) stated in Table 4.1. Then

$$[Y_0 \mid \boldsymbol{Y}^n] \sim T_1\left(\nu_i, \mu_i, \sigma_i^2\right), \tag{4.1.14}$$

where

$$\nu_i = \begin{cases} n + \nu_0\,, & i = (1) \\ n, & i = (2) \\ n - p + \nu_0, & i = (3) \\ n - p, & i = (4), \end{cases}$$

$$\mu_i = \mu_i(\boldsymbol{x}_0) = \begin{cases} \boldsymbol{f}_0^\top \boldsymbol{\mu}_{\boldsymbol{\beta}|n} + \boldsymbol{r}_0^\top \boldsymbol{R}^{-1}(\boldsymbol{y}^n - \boldsymbol{F}\boldsymbol{\mu}_{\boldsymbol{\beta}|n}), & i = (1) \text{ or } (2) \\ \boldsymbol{f}_0^\top \widehat{\boldsymbol{\beta}} + \boldsymbol{r}_0^\top \boldsymbol{R}^{-1}(\boldsymbol{y}^n - \boldsymbol{F}\widehat{\boldsymbol{\beta}}), & i = (3) \text{ or } (4) \end{cases}$$

with $\boldsymbol{\mu}_{\boldsymbol{\beta}|n} = \left(\boldsymbol{F}^\top \boldsymbol{R}^{-1}\boldsymbol{F} + \boldsymbol{V}_0^{-1}\right)^{-1} \left(\boldsymbol{F}^\top \boldsymbol{R}^{-1}\boldsymbol{y}^n + \boldsymbol{V}_0^{-1}\boldsymbol{b}_0\right)$,
$\widehat{\boldsymbol{\beta}} = \left(\boldsymbol{F}^\top \boldsymbol{R}^{-1}\boldsymbol{F}\right)^{-1} \left(\boldsymbol{F}^\top \boldsymbol{R}^{-1}\boldsymbol{y}^n\right)$, and

$$\sigma_i^2 = \sigma_i^2(\boldsymbol{x}_0) = \frac{Q_i^2}{\nu_i}\left\{1 - (\boldsymbol{f}_0^\top, \boldsymbol{r}_0^\top)\begin{bmatrix} \boldsymbol{V}_i & \boldsymbol{F}^\top \\ \boldsymbol{F} & \boldsymbol{R} \end{bmatrix}^{-1} \begin{pmatrix} \boldsymbol{f}_0 \\ \boldsymbol{r}_0 \end{pmatrix}\right\} \tag{4.1.15}$$

for $i = (1), \ldots, (4)$, where

$$\boldsymbol{V}_i = \begin{cases} -\boldsymbol{V}_0^{-1}, & i = (1) \text{ or } (2) \\ \boldsymbol{0}, & i = (3) \text{ or } (4) \end{cases}$$

and

$$Q_i^2 = \begin{cases} c_0 + Q_2^2, & i = (1) \\ Q_4^2 + \left(\boldsymbol{b}_0 - \widehat{\boldsymbol{\beta}}\right)^\top \left(\boldsymbol{V}_0 + [\boldsymbol{F}^\top \boldsymbol{R}^{-1}\boldsymbol{F}]^{-1}\right)^{-1} \left(\boldsymbol{b}_0 - \widehat{\boldsymbol{\beta}}\right), & i = (2) \\ c_0 + Q_4^2, & i = (3) \\ \boldsymbol{y}^{n\top}\left[\boldsymbol{R}^{-1} - \boldsymbol{R}^{-1}\boldsymbol{F}\left(\boldsymbol{F}^\top \boldsymbol{R}^{-1}\boldsymbol{F}\right)^{-1} \boldsymbol{F}^\top \boldsymbol{R}^{-1}\right]\boldsymbol{y}^n, & i = (4). \end{cases}$$

The formulas above for the degrees of freedom, location shift, and scale factor in the predictive t distribution all have very intuitive interpretations.

The base value for the degrees of freedom ν_i is $n-p$, which is augmented by p additional degrees of freedom when the prior for $\boldsymbol{\beta}$ is informative (cases (1) and (2)), and ν_0 additional degrees of freedom when the prior for σ_Z^2 is informative (cases (1) and (3)). For example, the degrees of freedom for case (4), with both components non-informative, is $n-p$ with no additions; the degrees of freedom for (1), with both components informative, is $n+\nu_0 = (n-p)+p+\nu_0$, corresponding to two incremental prior sources.

The location shift μ_i is precisely the same centering value as in Theorem 4.1.1 for the case of known σ_Z^2, either (4.1.4) or (4.1.8), depending on whether the informative or non-informative choice of prior is made for $[\boldsymbol{\beta} \mid \sigma_Z^2]$, respectively. In particular, the non-informative prior for $\boldsymbol{\beta}$ gives the BLUP (3.3.4).

The scale factor $\sigma_i^2(\boldsymbol{x}_0)$ in (4.1.15) is an estimate of the scale factor $\sigma_{0|n}^2(\boldsymbol{x}_0)$ in (4.1.6) of Theorem 4.1.1. The term in braces multiplying σ_Z^2 in (4.1.6) is the same as the term in braces in (4.1.15) after observing that $\tau^2 = \sigma_Z^2$ in Table 4.1. The remaining term in (4.1.15), Q_i^2/ν_i, is an estimate of σ_Z^2 in (4.1.6). The quadratic form, Q_i^2, pools information about σ_Z^2 from the conditional distribution of $\boldsymbol{Y}^n$ given σ_Z^2 with information from the prior of σ_Z^2 (when the latter is available). The scale factor $\sigma_i^2(\boldsymbol{x}_0)$ is zero when $\boldsymbol{x}_0$ is any of the training data points.

Theorem 4.1.2 is used to place pointwise prediction bands about $y(\boldsymbol{x}_0)$ by using the fact that, given $\boldsymbol{Y}^n$,

$$\frac{Y(\boldsymbol{x}_0) - \mu_i(\boldsymbol{x}_0)}{\sigma_i(\boldsymbol{x}_0)} \sim T_1(\nu_i, 0, 1)\,.$$

This gives

$$P\{Y(\boldsymbol{x}_0) \in \mu_i(\boldsymbol{x}_0) \pm \sigma_i(\boldsymbol{x}_0) t_{\nu_i}^{\alpha/2} | \boldsymbol{Y}^n\} = 1 - \alpha, \tag{4.1.16}$$

where $t_\nu^{\alpha/2}$ is the upper $\alpha/2$ critical point of the T_ν distribution (see Appendix A). When x_0 is real, $\mu_i(x_0) \pm \sigma_i(x_0)\, t_{\nu_i}^{\alpha/2}$ for $a < x_0 < b$ are pointwise $100(1-\alpha)\%$ prediction bands for $Y(x_0)$ at each $a < x_0 < b$.

Example 4.1 (Continued) Figure 4.2 plots the prediction bands corresponding to the BLUP when the predictive distribution is specified by the non-informative prior $[\boldsymbol{\beta}, \sigma_Z^2] \propto 1/\sigma_Z^2$ in Theorem 4.1.2. Notice that these bands have *zero width* at each of the true data points, as noted earlier. Prediction bands for any informative prior specification also have zero width at each of the true data points. ■

4.1.4 Prediction Distributions When Correlation Parameters Are Unknown

Subsections 4.1.2 and 4.1.3 assumed that the correlations among the observations are *known*, i.e., $\boldsymbol{R}$ and $\boldsymbol{r}_0$ are known. Now we assume that $y(\cdot)$

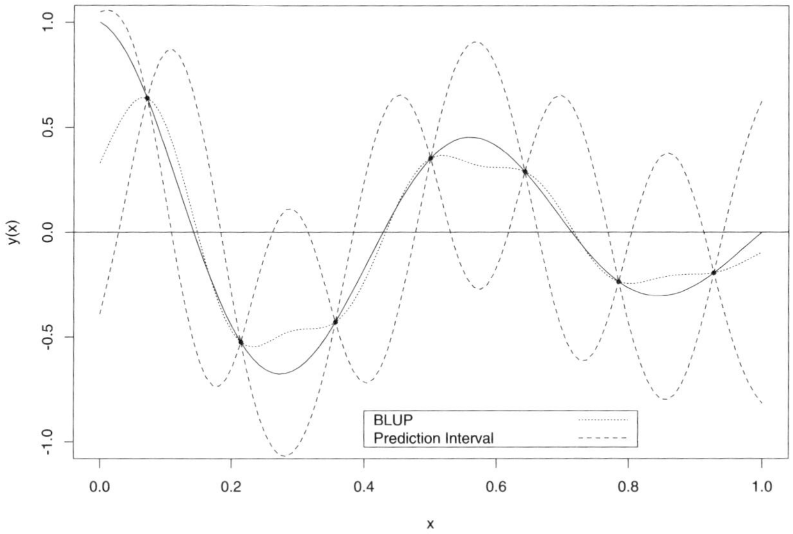

FIGURE 4.2. The BLUP and corresponding pointwise 95% prediction interval limits for $y(x)$ based on the non-informative prior of Theorem 4.1.2.

has a hierarchical Gaussian random field prior with parametric correlation function $R(\cdot \mid \boldsymbol{\psi})$ having *unknown* vector of parameters $\boldsymbol{\psi}$ (as introduced in Subsection 2.3.5 and previously considered in Subsection 3.3.2 for predictors). To facilitate the discussion below, suppose that the mean and variance of the normal predictive distribution in (4.1.3) and (4.1.8) are denoted by $\mu_{0|n}(\boldsymbol{x}_0) = \mu_{0|n}(\boldsymbol{x}_0|\boldsymbol{\psi})$ and $\sigma^2_{0|n}(\boldsymbol{x}_0) = \sigma^2_{0|n}(\boldsymbol{x}_0|\boldsymbol{\psi})$, where $\boldsymbol{\psi}$ was known in these earlier sections. Similarly, recall that the location and scale parameters of the predictive t distributions in (4.1.14) are denoted by $\mu_i(\boldsymbol{x}_0) = \mu_i(\boldsymbol{x}_0|\boldsymbol{\psi})$ and $\sigma^2_i(\boldsymbol{x}_0) = \sigma^2_i(\boldsymbol{x}_0|\boldsymbol{\psi})$, for $i \in \{(1), (2), (3), (4)\}$.

We consider two issues. The first is the assessment of the standard error of the plug-in predictor $\mu_{0|n}(\boldsymbol{x}_0|\widehat{\boldsymbol{\psi}})$ of $Y_0(\boldsymbol{x}_0)$ that is derived from $\mu_{0|n}(\boldsymbol{x}_0|\boldsymbol{\psi})$ by substituting $\widehat{\boldsymbol{\psi}}$, which is an estimator of $\boldsymbol{\psi}$ that might be the MLE or REML. This question is implicitly stated from the frequentist viewpoint. The second issue is Bayesian; we describe the Bayesian approach to uncertainty in $\boldsymbol{\psi}$ which is to model it by a prior distribution.

When $\boldsymbol{\psi}$ is *known*, recall that $\sigma^2_{0|n}(\boldsymbol{x}_0|\boldsymbol{\psi})$ is the MSPE of $\mu_{0|n}(\boldsymbol{x}_0|\boldsymbol{\psi})$. This suggests estimating the MSPE of $\mu_{0|n}(\boldsymbol{x}_0|\widehat{\boldsymbol{\psi}})$ by the plug-in MSPE $\sigma^2_{0|n}(\boldsymbol{x}_0|\widehat{\boldsymbol{\psi}})$. The correct expression for the MSPE of $\mu_{0|n}(\boldsymbol{x}_0|\widehat{\boldsymbol{\psi}})$ is

$$\text{MSPE}(\mu_{0|n}(\boldsymbol{x}_0|\widehat{\boldsymbol{\psi}}), \boldsymbol{\psi}) = E_{\boldsymbol{\psi}} \left\{ \left(\mu_{0|n}(\boldsymbol{x}_0|\widehat{\boldsymbol{\psi}}) - Y(\boldsymbol{x}_0) \right)^2 \right\}. \qquad (4.1.17)$$

Zimmerman and Cressie (1992) show that when the underlying surface is generated by a Gaussian random function,

$$\sigma^2_{0|n}(\boldsymbol{x}_0|\widehat{\boldsymbol{\psi}}) \le \text{MSPE}(\mu_{0|n}(\boldsymbol{x}_0|\widehat{\boldsymbol{\psi}}), \boldsymbol{\psi}) \tag{4.1.18}$$

under mild conditions so that $\sigma^2_{0|n}(\boldsymbol{x}_0|\widehat{\boldsymbol{\psi}})$ *underestimates* the true variance of the plug-in predictor. The amount of the underestimate is most severe when the underlying Gaussian random function has weak correlation. Zimmerman and Cressie (1992) propose a correction to $\sigma^2_{0|n}(\boldsymbol{x}_0|\widehat{\boldsymbol{\psi}})$ which provides a more nearly unbiased estimator of $\text{MSPE}(\mu_{0|n}(\boldsymbol{x}_0|\widehat{\boldsymbol{\psi}}), \boldsymbol{\psi})$. Nevertheless, $\sigma^2_{0|n}(\boldsymbol{x}_0|\widehat{\boldsymbol{\psi}})$ continues to be used for assessing the MSPE of $\mu_{0|n}(\boldsymbol{x}_0|\widehat{\boldsymbol{\psi}})$ because the amount by which it underestimates (4.1.17) has been shown to be asymptotically negligible for several models (for general linear models by Prasad and Rao (1990) and for time series models by Fuller and Hasza (1981)), and because of the lack of a compelling alternative that has demonstrably better small-sample properties.

An alternative viewpoint that accounts for uncertainty in $\boldsymbol{\psi}$ is to compute the mean squared prediction error based on the posterior distribution $[Y_0|\boldsymbol{Y}^n]$ (termed the "fully Bayesian approach" by some authors). We sketch how this is accomplished, at least in principle.

Assume that, in addition to $\boldsymbol{\beta}$ and σ^2_Z, knowledge about $\boldsymbol{\psi}$ is summarized in a 2nd stage $\boldsymbol{\psi}$ prior distribution. Often it will be reasonable to assume that the location and scale parameters, $\boldsymbol{\beta}$ and σ^2_Z, respectively, are independent of the correlation information so that the prior for the ensemble $[\boldsymbol{\beta}, \sigma^2_Z, \boldsymbol{\psi}]$ satisfies

$$[\boldsymbol{\beta}, \sigma^2_Z, \boldsymbol{\psi}] = [\boldsymbol{\beta}, \sigma^2_Z][\boldsymbol{\psi}].$$

For example, the non-informative prior of Theorem 4.1.2

$$[\boldsymbol{\beta}, \sigma^2_Z] = \frac{1}{\sigma^2_Z}$$

leads to the joint

$$[\boldsymbol{\beta}, \sigma^2_Z, \boldsymbol{\psi}] = \frac{1}{\sigma^2_Z}[\boldsymbol{\psi}].$$

In this case it is useful to regard the t posterior distributions that were stated in Theorem 4.1.2 as conditional on $\boldsymbol{\psi}$ and indicated by the notation $[Y_0|\boldsymbol{Y}^n, \boldsymbol{\psi}]$.

The required posterior distribution can be derived from

$$\begin{aligned} [Y(\boldsymbol{x}_0)|\boldsymbol{Y}^n] &= \int [Y(\boldsymbol{x}_0), \boldsymbol{\psi}|\boldsymbol{Y}^n]\, d\boldsymbol{\psi} \\ &= \int [Y(\boldsymbol{x}_0)|\boldsymbol{Y}^n, \boldsymbol{\psi}]\, [\boldsymbol{\psi}|\boldsymbol{Y}^n]\, d\boldsymbol{\psi} \end{aligned} \tag{4.1.19}$$

(see however the warning on page 69). The integration (4.1.19) can be prohibitive. For example, using the power exponential family with input-variable-specific scale and power parameters, the dimension of ψ is $2\times$ (number of inputs); ψ would be of dimension 12 for a six-dimensional input. Often, the posterior $[\psi|\boldsymbol{Y}^n]$ can be obtained from

$$[\psi|\boldsymbol{Y}^n] = \int \left[\boldsymbol{\beta}, \sigma_Z^2, \psi|\boldsymbol{Y}^n\right] d\boldsymbol{\beta}\; d\sigma_Z^2, \tag{4.1.20}$$

where the integrand in (4.1.20) is determined from

$$\left[\boldsymbol{\beta}, \sigma_Z^2, \psi|\boldsymbol{Y}^n\right] \propto \left[\boldsymbol{Y}^n|\boldsymbol{\beta}, \sigma_Z^2, \psi\right] \left[\boldsymbol{\beta}, \sigma_Z^2, \psi\right].$$

Equation (4.1.20) involves an integration of dimension equal to 1+ (the number of regressors), which is ordinarily less complicated than the integration (4.1.19) and can be carried out in closed form for "simple" priors.

In sum, one must both derive $[\psi|\boldsymbol{Y}^n]$ *and* carry out the typically high dimensional integration (4.1.19) in order to compute the required posterior distribution. Once the posterior is available, the Bayesian alternatives to $\mu_{0|n}(\boldsymbol{x}_0|\widehat{\psi})$ and $\sigma^2_{0|n}(\boldsymbol{x}_0|\widehat{\psi})$ are

$$E\left\{Y(\boldsymbol{x}_0)|\boldsymbol{Y}^n\right\}$$

and

$$\text{Var}\left\{Y(\boldsymbol{x}_0)|\boldsymbol{Y}^n\right\}.$$

Both the predictor and the associated assessment of accuracy account for uncertainty in ψ.

Handcock and Stein (1993) carried out the integrations in (4.1.19) for a specific two-input example using several regression models and isotropic correlation functions (power exponential and Matérn). As expected, they reported that for most cases that were studied, the Bayesian predictor and its standard errors gave wider confidence bands for the $Y(\boldsymbol{x}_0)$ than the plug-in predictors $\mu_{0|n}(\boldsymbol{x}_0|\widehat{\psi})$ and $\sigma^2_{0|n}(\boldsymbol{x}_0|\widehat{\psi})$. The plug-in predictor had particularly poor performance relative to the Bayes predictor when $\widehat{\psi}$ was determined by an eye-fit to the variogram associated with the correlation function.

We assess the magnitude of the underestimate of the plug-in MSPE estimator, which is given on the left-hand side of (4.1.18), by calculating the achieved coverage of the pointwise prediction intervals (4.1.16) having a given nominal level. The simulation results below consider only the four top predictors from Section 3.3; these top performing predictors were the EBLUPs based on the power exponential and Matérn correlation functions using either REML or MLE to estimate the unknown correlation parameters. In addition, only training data corresponding to the LHD and Sobol´ designs were used because the D-optimal design (assuming the cubic model)

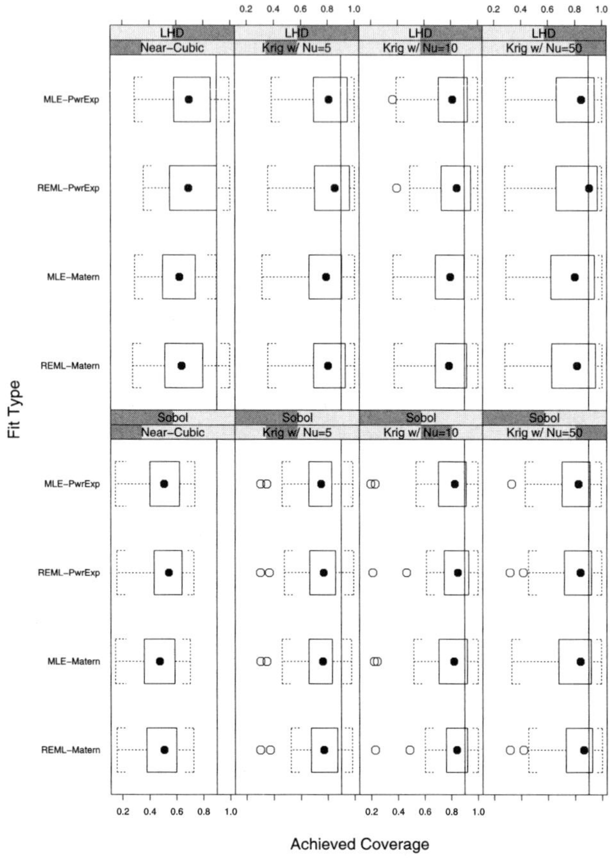

FIGURE 4.3. Box and whisker plots of the fifty proportions of the 625 equispaced grid of points in $[0, 1]^2$ that were covered by 90% nominal prediction intervals (4.1.16) classified by predictor, by experimental design, and type of surface.

tended to produce more biased predictions than the predictions based on training data using either the LHD or Sobol´ designs. For *each* of the 200 randomly selected surfaces on $[0, 1]^2$ that were described in the empirical study of Section 3.3, we computed the observed proportion of the 625 $\boldsymbol{x}_0$ points on $[0, 1]^2$ that were covered by the prediction interval (4.1.16) using $i = (4)$ (so that $\nu_i = n - 1$). This observed proportion was calculated for nominal 80%, 90%, 95%, and 99% prediction intervals. Figure 4.3 shows a trellis plot of a typical set of achieved coverages when $n = 16$, and the nominal coverage was 90%. Each box and whisker plot is based on the achieved coverages of the 50 randomly drawn surfaces from that combination of predictor, design, and surface.

The conclusions are as follows.

- Prediction intervals based on LHDs are slightly preferable to those based on Sobol′ designs, particularly for more irregular surfaces, i.e., surfaces with many local maxima and minima.
- All four EBLUPs produced nearly equivalent coverages, for each combination of the experimental design and source of random surface.
- For krigifier surfaces, the shortfall in the median coverage is 10% to 15%.

4.2 Prediction for Multiple Response Models

4.2.1 Introduction

This section derives predictors when several outputs are available from a computer experiment. One situation where this occurs is when several codes are available for computing the same response as, for example, when there are both "fast" (less accurate) and "slow" (more accurate) codes to compute an output. Such a hierarchy of codes is natural when, for example, finite element models of varying mesh sizes can be used to implement a mathematical model. Another setting in which multiple outputs occur is when there are competing responses (which might be expected to be *negatively* associated). The bivariate output example introduced in Section 1.2 that involved femoral stress shielding and femoral toggling (for given distributions of loading and trabecular bone elastic modulus) is such an example. A third example producing multiple outputs is when the code produces both $y(\cdot)$ and its derivatives; many engineering codes are of this type. The derivatives provide information about the $y(\cdot)$ surface. Example 4.2 considers this situation.

Initially, in Subsection 4.2.2, we describe several stochastic models for joint responses. Subsection 4.2.3 uses these models to describe, in principle, the optimal predictor for one of the several computed responses. Detailed examples are given to conclude the Section in 4.2.4. These multiple response models will be used again in Subsection 6.3.6, where an algorithm will be presented that locates a minimizing $\boldsymbol{x}_{\min}$ of $y_1(\boldsymbol{x})$ that satisfies feasibility constraints defined by $y_2(\cdot), \ldots, y_m(\cdot)$.

4.2.2 Modeling Multiple Outputs

This subsection describes a stochastic process model for multivariate outputs $y_1(\cdot)$, $y_2(\cdot)$, $\ldots$, $y_m(\cdot)$ having inputs $\boldsymbol{x} \in \mathcal{X} \subset \mathbb{R}^d$. As usual, we desire that this prior distribution embody uncertainties in the response $y_i(\cdot)$, $1 \leq i \leq m$.

Most of the models we consider have the form

$$Y_i(\boldsymbol{x}) = \boldsymbol{f}_i^\top(\boldsymbol{x})\boldsymbol{\beta}_i + Z_i(\boldsymbol{x}), \tag{4.2.1}$$

where the $Z_i(\cdot)$ are (marginally) mean zero stationary Gaussian stochastic processes with $Z_i(\cdot)$ having unknown variance $\sigma_i^2 > 0$, and correlation function $R_i(\cdot)$, $1 \leq i \leq m$. The linear model $\boldsymbol{f}_i^\top(\boldsymbol{x})\boldsymbol{\beta}_i$ represents the global mean of the Y_i process; here $\boldsymbol{f}_i(\cdot)$ is a $p_i \times 1$ vector of *known* regression functions and $\boldsymbol{\beta}_i \in \mathbb{R}^{p_i}$ is an *unknown* $p_i \times 1$ vector of regression parameters. We let $\boldsymbol{\beta} = (\boldsymbol{\beta}_1^\top, \ldots, \boldsymbol{\beta}_m^\top)^\top$ denote the vector of $p = \sum_1^m p_i$ regression parameters. Process stationarity of $Z_i(\boldsymbol{x})$ implies that the correlation between $Z_i(\boldsymbol{x}_1)$ and $Z_i(\boldsymbol{x}_2)$ depends only on $\boldsymbol{x}_1 - \boldsymbol{x}_2$. This model is completed upon specification of a *joint covariance structure* of the $\{Z_i(\cdot)\}$. In the subsequent general discussion, we assume $\text{Cov}\{Z_i(\boldsymbol{x}_1), Z_j(\boldsymbol{x}_2)\} = \sigma_i\sigma_j R_{ij}(\boldsymbol{x}_1 - \boldsymbol{x}_2)$, where $R_{ij}(\cdot)$ is the *cross-correlation function* of $Z_i(\cdot)$ and $Z_j(\cdot)$.

The choices of correlation and cross-correlation functions are complicated by the fact that they must "dovetail" to provide a valid overall correlation structure when the multiple responses are considered together. This means that for any choice of input sites $\boldsymbol{x}_\ell^i$, $1 \leq i \leq m$, and $1 \leq \ell \leq n_i$, at which the $Z_i(\cdot)$ are evaluated, the $(\sum_{i=1}^m n_i) \times 1$ multivariate normally distributed random vector $(Z_1(\boldsymbol{x}_1^1), \ldots, Z_1(\boldsymbol{x}_{n_1}^1), Z_2(\boldsymbol{x}_1^2), \ldots, Z_m(\boldsymbol{x}_{n_m}^m))^\top$ must have a positive definite (or at least nonnegative definite) covariance matrix.

Perhaps the most important method of assuring the validity of a multiple response model is to construct the $Z_i(\cdot)$ from a set of elementary (building block) processes; usually these building block processes are mutually independent. Such a model is used, for example, by Kennedy and O'Hagan (2000), who modeled the outputs of multi-level computer codes using a spatial autocorrelation structure. Let $Y_i(\boldsymbol{x})$ denote the prior for the ith code level ($i = 1, \ldots, m$). Let $i = m$ denote the top-level code, where *smaller* values of i indicate successively *less* complex code; our interest is typically in predicting the top-level output. One spatial autoregressive model for such multiple outputs is given by

$$Y_i(\boldsymbol{x}) = \rho_{i-1} Y_{i-1}(\boldsymbol{x}) + \delta_i(\boldsymbol{x}), \qquad i = 2, \ldots, m, \tag{4.2.2}$$

where $\delta_i(\cdot)$ is a process independent of $Y_1(\cdot)$, ..., $Y_{i-1}(\cdot)$. The output for each successive higher level code i at $\boldsymbol{x}$ is related to the output of the less precise code $i-1$ at $\boldsymbol{x}$ plus the refinement $\delta_i(\boldsymbol{x})$. This model can be embellished by allowing a separate regression for each stage of the model.

Under the assumption that $Y_i(\cdot)$ is stationary for each code i, this model is implied by

$$\text{Cov}\{\, Y_i(\boldsymbol{x}), Y_{i-1}(\boldsymbol{w}) \,|\, Y_{i-1}(\boldsymbol{x}) \,\} = 0, \qquad \text{for all} \quad \boldsymbol{w} \neq \boldsymbol{x},$$

which means that no additional second-order knowledge of code i at $\boldsymbol{x}$ can be obtained from the lower-level code $i-1$ at $\boldsymbol{w} \neq \boldsymbol{x}$ if the value of code

$i - 1$ at $\boldsymbol{x}$ is known. Therefore, the spatial autoregressive model can be interpreted as imposing a Markov property on the hierarchy of codes.

While the spatial autoregressive structure (4.2.2) can be employed as a modeling assumption in constrained optimization problems, there is no natural hierarchy of computer codes in such applications. A more reasonable model, which embodies the autoregressive spirit, is as follows. Let the output $y_1(\cdot)$ be the objective function and the outputs $y_2(\cdot), \ldots, y_{m+1}(\cdot)$ be m constraint functions. The model (4.2.2) is modified so that the constraint functions are each separately related to the objective function,

$$Y_i(\boldsymbol{x}) = \rho_i Y_1(\boldsymbol{x}) + \delta_i(\boldsymbol{x}), \qquad i = 2, \ldots, m+1, \tag{4.2.3}$$

where $\delta_i(\cdot)$ is independent of $Y_1(\cdot)$. This model says that each constraint function is associated with the objective function (positively or negatively) plus a refinement.

Constructive approaches have also been used to form models in the environmental sciences (see Ver Hoef and Barry (1998) or Higdon (1998b)). Such models for observed spatial data typically include an unknown smooth surface plus a random measurement error ("nugget effect"). As an example, Ver Hoef and Barry (1998) modeled *observed* spatial processes to be moving averages over white noise processes constructed by the following two-stage procedure. Let $W_0(\cdot)$, $W_1(\cdot)$, ..., $W_m(\cdot)$ denote mutually independent, mean zero, white noise processes. For $1 \le i \le m$, set

$$Z_i(\boldsymbol{w}) = \sqrt{1 - \rho_i^2}\, W_i(\boldsymbol{w}) + \rho_i W_0(\boldsymbol{w} - \boldsymbol{\Delta}_i),$$

where $-1 \le \rho_i \le 1$ and the parameter vectors $\boldsymbol{\Delta}_i$ represent spatial shifts in the cross-correlation function between two Z processes. Calculation gives

$$\mathrm{Cor}\{Z_{i_1}(\boldsymbol{w} + \boldsymbol{\Delta}_{i_1}), Z_{i_2}(\boldsymbol{w} + \boldsymbol{\Delta}_{i_2})\} = \rho_{i_1}\rho_{i_2},$$

so that nonzero cross-correlations are restricted to inputs separated spatially by the vector $\boldsymbol{\Delta}_{i_1} - \boldsymbol{\Delta}_{i_2}$. The model for the observed data are the integrated $Z_i(\cdot)$ processes with respect to a square integrable moving average function $f_i(\cdot \,|\, \boldsymbol{\theta}_i)$ plus a nugget (measurement error) effect and nonzero mean μ_i

$$Y_i(\boldsymbol{x}) = \int f_i(\boldsymbol{w} - \boldsymbol{x} \,|\, \boldsymbol{\theta}_i)\, Z_i(\boldsymbol{w})\, d\boldsymbol{w} + \nu_i U_i(\boldsymbol{x}) + \mu_i,$$

where $U_i(\cdot)$ is a zero mean, unit variance white noise process independent of $U_j(\cdot)$ for $j \ne i$. The square integrability condition on the moving average functions ensures that each $Y_i(\cdot)$ is second-order stationary. Ver Hoef and Barry (1998) found that it is possible to reproduce many commonly used variogram models in one dimension with this type of moving average construction.

Lastly, we turn attention to a model that is a useful prior for describing the multiple outputs of a code that produces both $y(\cdot)$ and the first partial derivatives of $y(\cdot)$. Suppose that the output from the code is $y(\cdot)$ for $\boldsymbol{x} \in \mathcal{X} \subset \mathbb{R}^d$ and that $y(\cdot)$ has first partial derivatives of all orders. Let

$$y^{(j)}(\boldsymbol{x}) = \partial y(\boldsymbol{x})/\partial x_j$$

denote the jth partial derivative of $y(\cdot)$ for $1 \leq j \leq d$. In the general notation introduced above, there are a total of $m = 1 + d$ outputs and $y_1(\cdot) = y(\cdot)$, $y_2(\cdot) = y^{(1)}(\cdot)$, …, $y_{1+m}(\cdot) = y^{(m)}(\cdot)$.

Morris et al. (1993) and Mitchell et al. (1994) consider a model for this setting in which the prior information about $y(\boldsymbol{x})$ is specified by a Gaussian process $Y(\cdot)$ and the prior about the partial derivatives $y^{(j)}(\boldsymbol{x})$ is obtained by considering the "derivative" processes of $Y(\cdot)$. Suppose that the prior for $y(\boldsymbol{x})$ is

$$Y(\boldsymbol{x}) = \boldsymbol{f}^\top(\boldsymbol{x})\boldsymbol{\beta} + Z(\boldsymbol{x}), \tag{4.2.4}$$

where each component of $\boldsymbol{f}^\top(\boldsymbol{x}) = (f_1(\boldsymbol{x}), \ldots, f_p(\boldsymbol{x}))$ has first partial derivatives with respect to all components x_j, $1 \leq j \leq d$, and $Z(\boldsymbol{x})$ is a stationary Gaussian process with zero mean, variance σ_Z^2, and product power exponential correlation function

$$R(\boldsymbol{w}) = \prod_{j=1}^{d} \exp\{-\theta_j \, w_j^2\}. \tag{4.2.5}$$

A natural prior for the partial derivative $y^{(j)}(\boldsymbol{x})$ is

$$Y^{(j)}(\boldsymbol{x}) = \lim_{h \to 0} \frac{Y(x_1, \ldots, x_{j-1}, x_j + h, x_{j+1}, \ldots x_d) - Y(\boldsymbol{x})}{h},$$

which exists under differentiability conditions for $R(\cdot)$ that are satisfied for the product power exponential correlation function (Parzen (1962)).

The "partial derivative" process $Y^{(j)}(\boldsymbol{x})$ has mean equal to $\frac{\partial \boldsymbol{f}^\top(\boldsymbol{x})\boldsymbol{\beta}}{\partial x_j} = \sum_{\ell=1}^{p} \beta_\ell \partial f_\ell(\boldsymbol{x})/\partial x_j$. We also require formulas for the covariances between the different partial derivative processes as well as between $Y(\boldsymbol{x})$ and each $Y^{(j)}(\boldsymbol{x})$. In general, when the $Y(\cdot)$ process need not be stationary, but has covariance function $\text{Cov}\{Y(\boldsymbol{x}^1), Y_{(j)}(\boldsymbol{x}^2)\} = R(\boldsymbol{x}^1, \boldsymbol{x}^2)$, the required formulas are

$$\text{Cov}\{Y(\boldsymbol{x}^1), Y^{(j)}(\boldsymbol{x}^2)\} = \sigma_Z^2 \frac{\partial R(\boldsymbol{x}^1, \boldsymbol{x}^2)}{\partial \boldsymbol{x}_j^2} \tag{4.2.6}$$

for $1 \leq j \leq d$ and

$$\text{Cov}\{Y^{(i)}(\boldsymbol{x}^1), Y^{(j)}(\boldsymbol{x}^2)\} = \sigma_Z^2 \frac{\partial^2 R(\boldsymbol{x}^1, \boldsymbol{x}^2)}{\partial \boldsymbol{x}_i^1 \partial \boldsymbol{x}_j^2} \tag{4.2.7}$$

for $1 \leq i \leq j \leq d$ (Morris et al. (1993)).

4.2.3 Optimal Predictors for Multiple Outputs

For sake of definiteness, let us fix attention on the problem of predicting $y_1(\cdot)$ at the (new) input site, $\boldsymbol{x}_0$, given output from all m codes, where each code is evaluated at its own unique set of training sites. Let $\boldsymbol{x}^i_\ell$, $1 \le \ell \le n_i$, denote the set of training sites for $y_i(\cdot)$, $1 \le i \le m$.

The best MSPE predictor based on this training data is, in principle, obtained from the results of Section 3.2 and the predictive distribution is given in Section 4.1. It is primarily a bookkeeping problem to set up the proper identifications in the notation of these earlier sections. The best MSPE predictor of $y_1(\boldsymbol{x}_0)$ is

$$\widehat{Y}_1(\boldsymbol{x}_0) = E\left\{Y_0 \,|\, \boldsymbol{Y}^{n_1}_1 = \boldsymbol{y}^{n_1}_1, \ldots, \boldsymbol{Y}^{n_m}_m = \boldsymbol{y}^{n_m}_m\right\}, \tag{4.2.8}$$

where $Y_0 = Y_1(\boldsymbol{x}_0)$, $\boldsymbol{Y}^{n_i}_i = (Y_i(\boldsymbol{x}^i_1), \ldots, Y_i(\boldsymbol{x}^i_{n_i}))$, and $\boldsymbol{y}^{n_i}_i$ is its observed value for $1 \le i \le m$. *The explicit formula for this conditional expectation depends on the joint distribution of* $(Y_1(\boldsymbol{x}_0), \boldsymbol{Y}^{n_1}_1, \ldots, \boldsymbol{Y}^{n_m}_m)$. The simplest cases for which one can derive a specific formula for (4.2.8) are Gaussian models. We first consider the one-stage Gaussian model (4.2.9) and later several two-stage models.

For the Gaussian model, the joint distribution of $(Y_0, \boldsymbol{Y}^{n_1}_1, \ldots, \boldsymbol{Y}^{n_m}_m)$ is the multivariate normal distribution

$$\begin{pmatrix} Y_0 \\ \boldsymbol{Y}^{n_1}_1 \\ \vdots \\ \boldsymbol{Y}^{n_m}_m \end{pmatrix} \sim N_{1+\sum_{i=1}^m n_i}\left[\boldsymbol{F}\,\boldsymbol{\beta}, \sigma^2_1\,\boldsymbol{\Sigma}\right], \tag{4.2.9}$$

where $\boldsymbol{F}$ and $\boldsymbol{\Sigma}$ are defined by

$$\begin{pmatrix} \boldsymbol{f}^\top_1(\boldsymbol{x}_0) & \cdots & \boldsymbol{0}_{1\times p_m} \\ \boldsymbol{F}_1 & \cdots & \boldsymbol{0}_{n_1\times p_m} \\ \vdots & \ddots & \vdots \\ \boldsymbol{0}_{n_m\times p_1} & \cdots & \boldsymbol{F}_m \end{pmatrix} \quad \text{and}$$

$$\begin{pmatrix} 1 & \boldsymbol{r}^\top_1 & \tau_2\,\boldsymbol{r}^\top_{12} & \cdots & \tau_m\,\boldsymbol{r}^\top_{1m} \\ \boldsymbol{r}_1 & \boldsymbol{R}_1 & \tau_2\,\boldsymbol{R}_{12} & \cdots & \tau_m\,\boldsymbol{R}_{1m} \\ \vdots & \vdots & \vdots & \ddots & \vdots \\ \tau_m\,\boldsymbol{r}_{1m} & \tau_m\,\boldsymbol{R}^\top_{1m} & \tau_m\,\boldsymbol{R}^\top_{2m} & \cdots & \tau^2_m\boldsymbol{R}_m \end{pmatrix},$$

respectively, where $\tau_i = \sigma_i/\sigma_1$, $2 \le i \le m$, and

- $\boldsymbol{f}_1(\boldsymbol{x}_0)$ is the $p_1 \times 1$ vector of regressors for $Y_1(\cdot)$ at $\boldsymbol{x}_0$,
- $\boldsymbol{F}_i = (\boldsymbol{f}^\top_i(\boldsymbol{x}^i_\ell))$ is the $n_i \times p_i$ matrix of regressors for the n_i inputs, where $y_i(\cdot)$ is evaluated for $1 \le i \le m$, and $1 \le \ell \le n_i$,

- $\boldsymbol{\beta} = (\boldsymbol{\beta}_1^\top, \ldots, \boldsymbol{\beta}_m^\top)^\top$, where $\boldsymbol{\beta}_i$ is the $p_i \times 1$ vector of regression coefficients associated with $\boldsymbol{Y}_i^{n_i}$, $1 \le i \le m$,

- $\boldsymbol{R}_i$ is the $n_i \times n_i$ matrix of correlations among the elements of $\boldsymbol{Y}_i^{n_i}$, $1 \le i \le m$,

- $\boldsymbol{r}_1 = (R_1(\boldsymbol{x}_0 - \boldsymbol{x}_1^1), \ldots, R_1(\boldsymbol{x}_0 - \boldsymbol{x}_{n_1}^1))^\top$ is the $n_1 \times 1$ vector of correlations of $Y_1(\boldsymbol{x}_0)$ with $\boldsymbol{Y}_1^{n_1}$,

- $\boldsymbol{r}_{1i} = (R_{1i}(\boldsymbol{x}_0 - \boldsymbol{x}_1^i), \ldots, R_{1i}(\boldsymbol{x}_0 - \boldsymbol{x}_{n_i}^i))^\top$ is the $n_i \times 1$ vector of correlations of $Y_1(\boldsymbol{x}_0)$ with $\boldsymbol{Y}_i^{n_i}$, $2 \le i \le m$,

- $\boldsymbol{R}_{ij}$ is the $n_i \times n_j$ matrix of correlations between $\boldsymbol{Y}_i^{n_i}$ and $\boldsymbol{Y}_j^{n_j}$, $1 \le i < j \le m$.

By Section 3.2, the conditional expectation (4.2.8) is given by

$$\boldsymbol{f}_0^\top \boldsymbol{\beta} + \boldsymbol{r}_0^\top \boldsymbol{R}^{-1} (\boldsymbol{Y}^n - \boldsymbol{F}\boldsymbol{\beta}) \tag{4.2.10}$$

with the identifications $\boldsymbol{f}_0^\top = \left(\boldsymbol{f}_1^\top(\boldsymbol{x}_0), \boldsymbol{0}_{1\times(p-p_1)}\right)$, $\boldsymbol{r}_0^\top = (\boldsymbol{r}_1^\top, \tau_2\, \boldsymbol{r}_{12}^\top, \cdots, \tau_m\, \boldsymbol{r}_{1m}^\top)$, $\boldsymbol{R}$ is the bottom right $(\sum_{i=1}^m n_i) \times (\sum_{i=1}^m n_i)$ submatrix of $\boldsymbol{\Sigma}$, $\boldsymbol{Y}^n$ is the $(\sum_{i=1}^m n_i) \times 1$ vector of observed outputs, and $\boldsymbol{F}$ and $\boldsymbol{\beta}$ are as in (4.2.9). In practice, the formula (4.2.10) is seldom likely to be useful because it requires knowledge of marginal correlation functions, joint correlation functions, *and* the ratio of all the process variances.

However, *empirical versions* of this predictor *are* of practical use. The MLE- and REML- EBLUPs described in Subsection 3.3.2 both assume that each of the correlation matrices $\boldsymbol{R}_i$, $1 \le i \le m$, and cross-correlation matrices $\boldsymbol{R}_{ij}$, $1 \le i < j \le m$, are known up to a finite vector of parameters. Suppose that $\boldsymbol{\psi}_i$ is the vector of unknown parameters for $R_i(\cdot)$, and $\boldsymbol{\psi}_{ij}$ is the unknown parameter vector for $R_{ij}(\cdot)$. Then the $\boldsymbol{\psi} = (\tau_2, \ldots, \tau_m, \boldsymbol{\psi}_1, \ldots, \boldsymbol{\psi}_m, \boldsymbol{\psi}_{12}, \ldots, \boldsymbol{\psi}_{m-1,m})^\top$ contains all the unknown parameters required to describe the correlations of $Y_1(\boldsymbol{x}_0)$ with the training data.

As sketched in Subsection 4.1.4, a fully Bayesian predictor for this setup puts a prior on $[\boldsymbol{\beta}, \sigma_1^2, \boldsymbol{\psi}]$ and uses the mean of the predictive distribution (4.2.8) as the desired predictor. For multiple response models it is even more difficult analytically and numerically to construct this predictor than in the single response case, and we have previously noted that the single response case can be very difficult indeed (Handcock and Stein (1993)).

As in Subsection 3.3.2, we use the predictor

$$\widehat{Y}_1(\boldsymbol{x}_0) = E\left\{Y_0 | \boldsymbol{Y}_1^{n_1} = \boldsymbol{y}_1^{n_1}, \ldots, \boldsymbol{Y}_m^{n_m} = \boldsymbol{y}_m^{n_m}, \widehat{\boldsymbol{\psi}}\right\},$$

where $\widehat{\boldsymbol{\psi}}$ is estimated from $(\boldsymbol{Y}_1^{n_1}, \ldots, \boldsymbol{Y}_m^{n_m})$ based on the Gaussian likelihood (or restricted likelihood) induced from (4.2.9).

As developed in Section 4.1, this predictor can be obtained by first considering a two-stage model in which ψ is *known*. Suppose that the conditional distribution of $(Y_0, \boldsymbol{Y}_1^{n_1}, \ldots, \boldsymbol{Y}_m^{n_m})$ given $(\boldsymbol{\beta}, \sigma_1^2)$ is (4.2.9), and the marginal distribution of $(\boldsymbol{\beta}, \sigma_1^2)$ is the (non-informative) prior

$$[\boldsymbol{\beta}, \sigma_1^2] \propto \frac{1}{\sigma_1^2}. \tag{4.2.11}$$

The predictor, $E\{Y_0 | \boldsymbol{Y}_1^{n_1} = \boldsymbol{y}_1^{n_1}, \ldots, \boldsymbol{Y}_m^{n_m} = \boldsymbol{y}_m^{n_m}, \psi\}$, corresponding to this two-stage model is

$$\widehat{Y}_1(\boldsymbol{x}_0) = \boldsymbol{f}_0^\top \widehat{\boldsymbol{\beta}} + \boldsymbol{r}_0^\top \boldsymbol{R}^{-1}(\boldsymbol{Y}^n - \boldsymbol{F}\widehat{\boldsymbol{\beta}}), \tag{4.2.12}$$

where $\boldsymbol{f}_0$, $\boldsymbol{r}_0$, $\boldsymbol{R}$, $\boldsymbol{Y}^n$, and $\boldsymbol{F}$ are described following Equation (4.2.10) and $\widehat{\boldsymbol{\beta}}$ is the generalized least squares estimator of $\boldsymbol{\beta}$ based on $\boldsymbol{Y}^n$. When ψ is *unknown*, we estimate ψ in (4.2.12) using MLE or REML to produce an EBLUP of $Y_1(\boldsymbol{x}_0)$.

4.2.4 Examples

Example 4.2 Consider a code having bivariate input $\boldsymbol{x} = (x_1, x_2)$ that also produces the partial derivates of $y(\cdot)$, $y^{(1)}(\boldsymbol{x}) = \partial y(\boldsymbol{x})/\partial x_1$, and $y^{(2)}(\boldsymbol{x}) = \partial y(\boldsymbol{x})/\partial x_2$. Assume the model (4.2.4) of Subsection 4.2.2 with product power exponential correlation function

$$R(h_1, h_2) = \exp\{-\theta_1\, h_1^2\} \times \exp\{-\theta_2\, h_2^2\}.$$

Let $\boldsymbol{x}^1 = (x_1^1, x_2^1)$ and $\boldsymbol{x}^2 = (x_1^2, x_2^2)$. Applying the formulas (4.2.6) and (4.2.7) gives the pairwise joint covariances of $Y(\cdot)$, $Y^{(1)}(\cdot)$, and $Y^{(2)}(\cdot)$:

$$\mathrm{Cov}\{Y(\boldsymbol{x}^1), Y^{(j)}(\boldsymbol{x}^2)\} = -2\,\theta_j\,(x_j^1 - x_j^2)\,\sigma_Z^2\, R(\boldsymbol{x}^1 - \boldsymbol{x}^2),$$

$$\mathrm{Cov}\{Y^{(j)}(\boldsymbol{x}^1), Y^{(j)}(\boldsymbol{x}^2)\} = \left(2\,\theta_j - 4\,\theta_j^2\,(x_j^1 - x_j^2)^2\right)\sigma_Z^2\, R(\boldsymbol{x}^1 - \boldsymbol{x}^2)$$

for $j = 1, 2$, and

$$\mathrm{Cov}\{Y^{(1)}(\boldsymbol{x}^1), Y^{(2)}(\boldsymbol{x}^2)\} = 4\,\theta_1\,\theta_2(x_1^1 - x_1^2)\,(x_2^1 - x_2^2)\,\sigma_Z^2 R(\boldsymbol{x}^1 - \boldsymbol{x}^2).$$

Using these covariance functions, the EBLUP based on $y(\boldsymbol{x})$, $y^{(1)}(\boldsymbol{x})$, and $y^{(2)}(\boldsymbol{x})$ can be computed from (4.2.12) with appropriate code to estimate the scale parameters (θ_1, θ_2) and to implement the predictor.

As a specific numerical example, let

$$y(x_1, x_2) = 2\,x_1^3\,x_2^2$$

on $[-1, 1]^2$, which is displayed in Figure 4.4. The cubic and quadratic characters of $y(\cdot)$ in x_1 and x_2, respectively, are clearly visible on this domain. The first partial derivatives of $y(\cdot)$ are

$$y^{(1)}(x_1, x_2) = \partial y(x_1, x_2)/\partial x_1 = 6\,x_1^2\,x_2^2$$

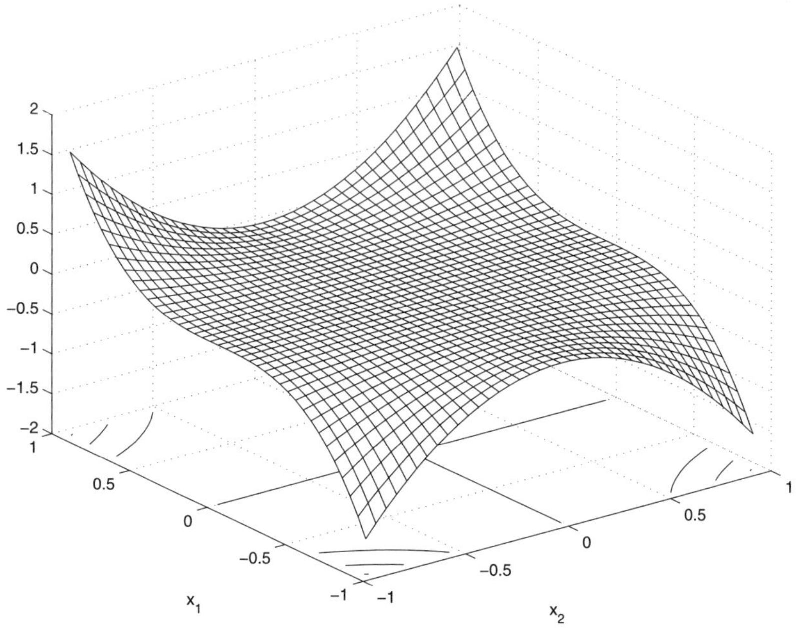

FIGURE 4.4. The function $y(x_1, x_2) = 2x_1^3 x_2^2$ on $[-1, 1]^2$.

and

$$y^{(2)}(x_1, x_2) = \partial y(x_1, x_2)/\partial x_2 = 4\ x_1^3\ x_2.$$

Consider the 14 point training data displayed in Figure 4.5. This set of locations has the intuitive feature that it allows us to "learn" about $y(\boldsymbol{x})$ and its partial derivatives over the the entire input space—the design is "space-filling." Space-filling designs are discussed in detail in Section 5.2.

We illustrate the benefit of adding the derivative information by predicting $y(\cdot)$ on the 1521 ($= 39^2$) grid of points $[-.95(.05).95]^2$ by first using the predictor of $y(\cdot)$ based on $y(\cdot)$ *alone*, and then by using the predictor of $y(\cdot)$ based on $\left(y(\cdot), y^{(1)}(\cdot), y^{(2)}(\cdot)\right)$. The regression function for $Y(\cdot)$ is taken to be constant, β_0. We measure the accuracy of the generic predictor $\widehat{y}(\cdot)$ of $y(\cdot)$ by its *empirical root mean squared prediction error* (ERMSPE) at the 1521 point test grid, which is defined to be

$$\text{ERMSPE}(\widehat{y}) = \sqrt{\frac{1}{1521}\sum_{i=1}^{1521}\left(y(\boldsymbol{x}_i) - \widehat{y}(\boldsymbol{x}_i)\right)^2}. \tag{4.2.13}$$

Table 4.2 summarizes the estimated model parameters and ERMSPEs for the two predictors. Figure 4.6 displays the predictor of $y(\cdot)$ based on the 14 point training set evaluations of $y(\cdot)$; this predictor has ERMSPE equal to 0.3180. Figure 4.7 plots the predictor of $y(\cdot)$ based on the 14

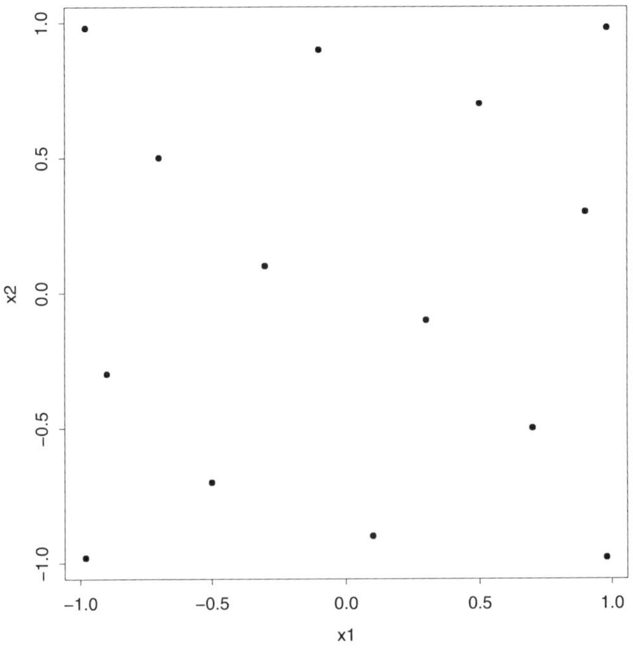

FIGURE 4.5. Fourteen point training design on $[-1, 1]^2$.

	Predictor Based On	
	$y(\cdot)$	$\left(y(\cdot), y^{(1)}(\cdot), y^{(2)}(\cdot)\right)$
$\widehat{\theta}_1$	0.7071	13.4264
$\widehat{\theta}_2$	10.9082	8.7060
$\widehat{\beta}_0$	0.00	0.00
ERMSPE	0.3180	0.2566

TABLE 4.2. Empirical RMSPEs at the 39^2 points on the grid $[-.95(.05).95]^2$ for the EBLUPs based on $y(\cdot)$ alone and $\left(y(\cdot), y^{(1)}(\cdot), y^{(2)}(\cdot)\right)$.

point training set evaluations of $\left(y(\cdot), y^{(1)}(\cdot), y^{(2)}(\cdot)\right)$; it has ERMSPE equal to 0.2566. The fit based on $y(\cdot)$ and its derivatives is *both* more visually appealing (compare prediction of the x_1 edges with those of the true surface in Figure 4.4) *and* has a 24% smaller ERMSPE than that based on $y(\cdot)$ alone. ■

Example 4.3 This example considers two responses that are defined on a common input space $\mathcal{X}$. The setting is that described by Kennedy and O'Hagan (2000), who considered *predicting* the outcome of a finite element

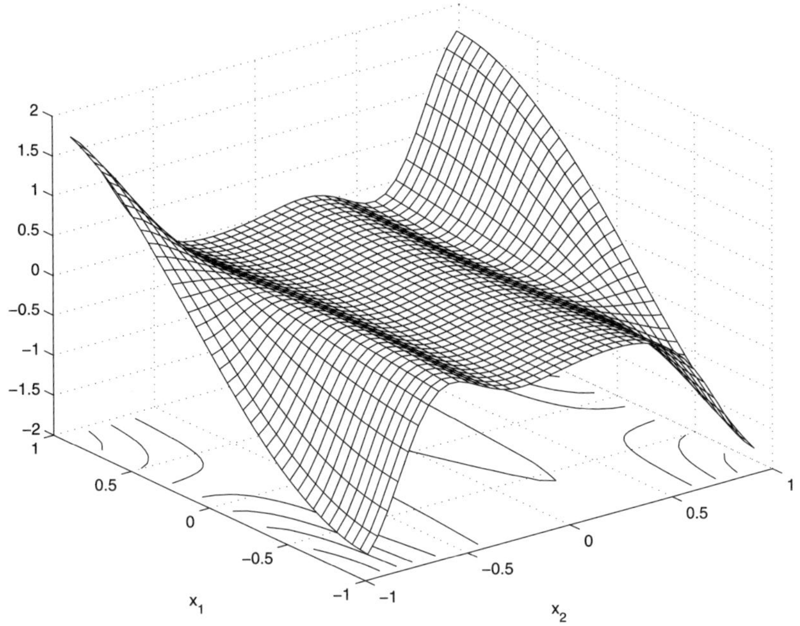

FIGURE 4.6. Prediction surface based on $y(\cdot)$ using the 14 point exploratory design and the power exponential correlation function.

code that uses a very fine grid based on output from both this fine grid code and from the output of a faster but less accurate code that is based on a coarser grid. In practice, there could be a hierarchy of m codes with each succeeding code more accurate than its predecessor.

To describe this situation more formally, suppose that $y_p(\cdot)$ represents the output from fast, but *poorer*, code and $y_g(\cdot)$ the output from the slow, but *good*, code. *Our goal is to predict the output of the good code at input site* $\boldsymbol{x}_0$, *i.e., to predict* $y_g(\boldsymbol{x}_0)$, *based on* $y_p(\boldsymbol{x}_1^p)$, $\ldots$, $y_p(\boldsymbol{x}_{n_p}^p)$ *and* $y_g(\boldsymbol{x}_1^g)$, $\ldots$, $y_g(\boldsymbol{x}_{n_g}^g)$.

Consider a multi-stage model for this setting in which the prior for $y_p(\cdot)$ is

$$Y_p(\boldsymbol{x}) = \boldsymbol{f}_p^\top(\boldsymbol{x})\boldsymbol{\beta}_p + W_p(\boldsymbol{x}), \tag{4.2.14}$$

where the regression function $\boldsymbol{f}_p^\top(\cdot)$ specifies the large-scale, nonstationary structure of $y_p(\cdot)$ and $W_p(\cdot)$ is a stationary Gaussian process that determines the local features of the code; $W_p(\cdot)$ is assumed to have zero mean, variance σ_p^2, and correlation function $R_p(\cdot)$. We use the notation $W_p(\cdot)$ rather than $Z_p(\cdot)$ to avoid confusion in the following discussion. A prior for

PErK Job for Example-4.3

```
NumberOfInputs = 2
CorrelationFamily1 = PowerExponential
CorrelationFamily2 = PowerExponential
RandomError = No
Tries = 5
LogLikelihoodTolerance = 1.e-2
SimplexTolerance = 1.e-2
RandomNumberSeed = 1472
CorrelationEstimation = REML
X1 < xg.7
Y1 < yg.7
RegressionModel1 < reg.constant
X2 < xp.11
Y2 < yp.11
RegressionModel2 < reg.constant
XPred < xg.new

Summary > 4.3.summary
RegressionModel > 4.3.beta
StochasticProcessModel > 4.3.corpar
```

$y_g(\cdot)$ data only		$y_g(\cdot)$ & $y_p(\cdot)$ data	
MLE	REML	MLE	REML
0.196	0.155	0.196	0.031

TABLE 4.3. Empirical root mean squared prediction errors of four $y_g(\cdot)$ predictors (MLE versus RMLE estimation of the correlation parameters; $y_g(\cdot)$ training data only versus both $y_g(\cdot)$ and $y_p(\cdot)$ training data).

Table 4.3 lists the empirical root mean squared prediction error based on a 100 point equispaced grid on $(0, 1)$ for the four $y_g(\cdot)$ predictors. The $y_p(\cdot)$ data adds essentially nothing to the estimation process when using MLE to estimate the correlation parameters. However, the situation is dramatically different when using REML. In this case, the $y_g(\cdot)$ predictor that uses *both* outputs is a great improvement over that based on the $y_g(\cdot)$ training data alone; the ERMSPE is improved five-fold by using the $y_p(\cdot)$ information. Figure 4.10 plots the target function $y_g(\cdot)$ and the predictors based on the REMLs of the correlation parameters. Clearly, the $y_p(\cdot)$ training data near the origin allows us to indirectly "see" the sharp negative slope in $y_g(\cdot)$ in this region. ■

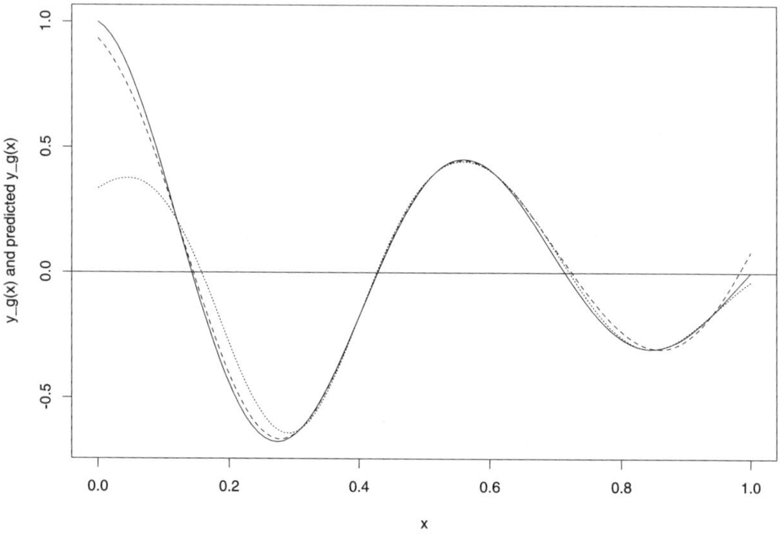

FIGURE 4.10. The target function $y_g(x)$ (solid), the EBLUP based on $y_g(x)$ (dotted), and the EBLUP based on $(y_p(x), y_g(x))$ (dashed) with correlation parameters estimated by REML.

4.3 Chapter Notes

Outline of Proofs of Theorems 4.1.1 and 4.1.2

The strategy for proving Theorems 4.1.1 and 4.1.2 is as follows. Suppose that $(Y_0, \boldsymbol{Y}^n)$ has a conditional distribution $[(Y_0, \boldsymbol{Y}^n) \mid \boldsymbol{\omega}]$ given the generic (vector) parameter $\boldsymbol{\omega}$. If the conditional distribution $[\boldsymbol{\omega} \mid \boldsymbol{Y}^n]$ can be determined, then

$$[(Y_0, \boldsymbol{\omega}) \mid \boldsymbol{Y}^n] = [Y_0 \mid \boldsymbol{Y}^n, \boldsymbol{\omega}] \times [\boldsymbol{\omega} \mid \boldsymbol{Y}^n];$$

integrating out $\boldsymbol{\omega}$ in the joint density $[(Y_0, \boldsymbol{\omega}) \mid \boldsymbol{Y}^n]$ gives the required predictive distribution.

Some Details of the Proof of Theorem 4.1.1

Carrying out the strategy above in the case of Theorem 4.1.1 is straightforward, in principle, although tedious in its details. Throughout, we suppress the dependence of the various conditional distributions on the known values σ_Z^2. The density of $[Y_0 \mid \boldsymbol{Y}^n]$ can be written as

$$\begin{aligned} f(y_0 \mid \boldsymbol{y}^n) &= \int_{I\!R^p} f(y_0, \boldsymbol{\beta} \mid \boldsymbol{y}^n)\, d\boldsymbol{\beta} \\ &= \int_{I\!R^p} f(y_0 \mid \boldsymbol{y}^n, \boldsymbol{\beta}) f(\boldsymbol{\beta} \mid \boldsymbol{y}^n)\, d\boldsymbol{\beta}\,. \end{aligned} \quad (4.3.1)$$

This calculation can be simplified by recognizing that we only need keep track of terms involving y_0 and $\boldsymbol{\beta}$; the former are required to state the functional form of $f(y_0 \mid \boldsymbol{y}^n)$ and the latter are required to carry out the integration (4.3.1). All other terms are part of the normalizing constant for $f(y_0 \mid \boldsymbol{y}^n)$.

The first term in the integrand of (4.3.1) follows immediately by applying Lemma B.1.2 concerning the conditional normal distribution so that

$$[Y_0 \mid \boldsymbol{Y}^n, \boldsymbol{\beta}] \sim N_1 \left[\mu_{0|n,\boldsymbol{\beta}}, \sigma^2_{0|n,\boldsymbol{\beta}}\right],$$

where

$$\mu_{0|n,\boldsymbol{\beta}} = \boldsymbol{f}_0^\top \boldsymbol{\beta} + \boldsymbol{r}_0^\top \boldsymbol{R}^{-1}(\boldsymbol{y}^n - \boldsymbol{F}\boldsymbol{\beta}) \text{ and } \sigma^2_{0|n,\boldsymbol{\beta}} = \sigma^2_Z(1 - \boldsymbol{r}_0^\top \boldsymbol{R}^{-1}\boldsymbol{r}_0).$$

The second density can be derived by observing that

$$\begin{aligned} f(\boldsymbol{\beta} \mid \boldsymbol{y}^n) &\propto f(\boldsymbol{y}^n \mid \boldsymbol{\beta}) \times f(\boldsymbol{\beta}) \\ &\propto \exp\left\{ -\frac{1}{2}(\boldsymbol{y}^n - \boldsymbol{F}\boldsymbol{\beta})^\top \frac{\boldsymbol{R}^{-1}}{\sigma^2_Z}(\boldsymbol{y}^n - \boldsymbol{F}\boldsymbol{\beta}) \right. \\ &\qquad \left. -\frac{1}{2}(\boldsymbol{\beta} - \boldsymbol{b}_0)^\top \frac{\boldsymbol{V}_0^{-1}}{\tau^2}(\boldsymbol{\beta} - \boldsymbol{b}_0) \right\}. \end{aligned}$$

Collecting terms that depend on $\boldsymbol{\beta}$ and grouping the others with the constant of proportionality gives

$$f(\boldsymbol{\beta} \mid \boldsymbol{y}^n) \propto \exp\left\{ -\frac{1}{2}\boldsymbol{\beta}^\top \boldsymbol{A}^{-1}\boldsymbol{\beta} + \boldsymbol{\nu}^\top \boldsymbol{\beta} \right\},$$

where $\boldsymbol{A}^{-1} = \left[\frac{\boldsymbol{F}^\top \boldsymbol{R}^{-1}\boldsymbol{F}}{\sigma^2_Z} + \frac{\boldsymbol{V}_0^{-1}}{\tau^2}\right]$ and $\boldsymbol{\nu} = \left[\frac{\boldsymbol{F}^\top \boldsymbol{R}^{-1}\boldsymbol{y}^n}{\sigma^2_Z} + \frac{\boldsymbol{V}_0^{-1}\boldsymbol{b}_0}{\tau^2}\right]$. By Lemma B.1.1,

$$\boldsymbol{\beta} \mid \boldsymbol{Y}^n \sim N_p\left[\boldsymbol{\mu}_{\boldsymbol{\beta}|n} \equiv \boldsymbol{A}\boldsymbol{\nu}, \boldsymbol{\Sigma}_{\boldsymbol{\beta}|n} \equiv \boldsymbol{A}\right],$$

where $\boldsymbol{\mu}_{\boldsymbol{\beta}|n}$ is defined by (4.1.5) and $\boldsymbol{\Sigma}_{\boldsymbol{\beta}|n} = \left[\frac{\boldsymbol{F}^\top \boldsymbol{R}^{-1}\boldsymbol{F}}{\sigma^2_Z} + \frac{\boldsymbol{V}_0^{-1}}{\tau^2}\right]^{-1}$.

Returning to (4.3.1) and ignoring constants of proportionality, we obtain

$$\begin{aligned} f(y_0 \mid \boldsymbol{y}^n) &\propto \int_{I\!R^p} \exp\left\{ -\frac{1}{2}\frac{(y_0 - \mu_{0|n,\boldsymbol{\beta}})^2}{\sigma^2_{0|n,\boldsymbol{\beta}}} \right. \\ &\qquad \left. -\frac{1}{2}(\boldsymbol{\beta} - \boldsymbol{\mu}_{\boldsymbol{\beta}|n})^\top \boldsymbol{\Sigma}^{-1}_{\boldsymbol{\beta}|n}(\boldsymbol{\beta} - \boldsymbol{\mu}_{\boldsymbol{\beta}|n}) \right\} d\boldsymbol{\beta} \\ &\propto \exp\left\{ -\frac{1}{2}\frac{(y_0 - \boldsymbol{r}_0^\top \boldsymbol{R}^{-1}\boldsymbol{y}^n)^2}{\sigma^2_{0|n,\boldsymbol{\beta}}} \right\} \\ &\qquad \times \int_{I\!R^p} \exp\left\{ -\frac{1}{2}\boldsymbol{\beta}^\top \boldsymbol{A}^{-1}\boldsymbol{\beta} + \boldsymbol{\nu}^\top \boldsymbol{\beta} \right\} d\boldsymbol{\beta}, \end{aligned}$$

which is again evaluated by applying Lemma B.1.1, where $\boldsymbol{h} = \boldsymbol{f}_0 - \boldsymbol{F}^\top \boldsymbol{R}^{-1}\boldsymbol{r}_0$, $\boldsymbol{A}^{-1} = \left(\frac{\boldsymbol{h}\boldsymbol{h}^\top}{\sigma^2_{0|n,\boldsymbol{\beta}}} + \boldsymbol{\Sigma}^{-1}_{\boldsymbol{\beta}|n} \right)$ in this application, and
$\boldsymbol{\nu} = \left(\boldsymbol{h} \left(y_0 - \boldsymbol{r}_0^\top \boldsymbol{R}^{-1} \boldsymbol{y}^n \right) / \sigma^2_{0|n,\boldsymbol{\beta}} + \boldsymbol{\Sigma}^{-1}_{\boldsymbol{\beta}|n} \boldsymbol{\mu}_{\boldsymbol{\beta}|n} \right)$.

Keeping track of the terms involving y_0, which are required to determine the form in which they enter the expression for the $f(y_0 \mid \boldsymbol{y}^n)$ density, and rearranging, we obtain that

$$f(y_0 \mid \boldsymbol{y}^n) = \exp\left\{ -\frac{1}{2a} y_0^2 + y_0 \times b \right\},$$

where

$$\begin{aligned} a &= \sigma^2_Z \left\{ 1 - \boldsymbol{r}_0^\top \boldsymbol{R}^{-1} \boldsymbol{r}_0 \right\} + \left(\boldsymbol{f}_0^\top - \boldsymbol{r}_0^\top \boldsymbol{R}^{-1} \boldsymbol{F} \right) \boldsymbol{\Sigma}_{\boldsymbol{\beta}|n} \left(\boldsymbol{f}_0 - \boldsymbol{F}^\top \boldsymbol{R}^{-1} \boldsymbol{r}_0 \right) \\ &= \sigma^2_Z \left\{ 1 - \boldsymbol{r}_0^\top \boldsymbol{R}^{-1} \boldsymbol{r}_0 + \boldsymbol{h}^\top \left[\boldsymbol{F}^\top \boldsymbol{R}^{-1} \boldsymbol{F} + \frac{\sigma^2_Z}{\tau^2} \boldsymbol{V}_0^{-1} \right] \boldsymbol{h} \right\} \end{aligned}$$

and

$$b = (\boldsymbol{f}_0^\top \boldsymbol{\mu}_{\boldsymbol{\beta}|n} + \boldsymbol{r}_0^\top \boldsymbol{R}^{-1} (\boldsymbol{y}^n - \boldsymbol{F} \boldsymbol{\mu}_{\boldsymbol{\beta}|n})) \times a^{-1}.$$

Solving for $\sigma^2_{0|n} = a$ and $\mu_{0|n} = a{\cdot}b$ gives the result of Theorem 4.1.1. □

Some Details of the Proof of Theorem 4.1.2

The proof of Theorem 4.1.2 is similar in spirit to that of Theorem 4.1.1 but is complicated by the presence of the extra parameter σ^2_Z. We calculate the distribution of $[Y_0 \mid \boldsymbol{Y}^n]$ from

$$[Y_0 \mid \boldsymbol{Y}^n] = \int [Y_0, \sigma^2_Z \mid \boldsymbol{Y}^n] \, d\sigma^2_Z.$$

The integrand is obtained from the product

$$[Y_0, \sigma^2_Z \mid \boldsymbol{Y}^n] = [Y_0 \mid \boldsymbol{Y}^n, \sigma^2_Z][\sigma^2_Z \mid \boldsymbol{Y}^n]. \tag{4.3.2}$$

The left-hand term in (4.3.2) is calculated from

$$[Y_0 \mid \boldsymbol{Y}^n, \sigma^2_Z] = \int [Y_0, \boldsymbol{\beta} \mid \boldsymbol{Y}^n, \sigma^2_Z] \, d\boldsymbol{\beta}$$

whose integrand can be obtained from

$$[Y_0, \boldsymbol{\beta} \mid \boldsymbol{Y}^n, \sigma^2_Z] = [Y_0 \mid \boldsymbol{Y}^n, \boldsymbol{\beta}, \sigma^2_Z][\boldsymbol{\beta} \mid \boldsymbol{Y}^n, \sigma^2_Z]. \tag{4.3.3}$$

The left-hand term of (4.3.3) is known and $[\boldsymbol{\beta} \mid \boldsymbol{Y}^n, \sigma^2_Z]$ can be deduced from

$$[\boldsymbol{\beta} \mid \boldsymbol{Y}^n, \sigma^2_Z] \propto [\boldsymbol{Y}^n \mid \boldsymbol{\beta}, \sigma^2_Z]. \tag{4.3.4}$$

The right-hand term in (4.3.2) can be obtained by noting

$$\begin{aligned} [\sigma_Z^2 \mid \boldsymbol{Y}^n] &\propto [\sigma_Z^2, \boldsymbol{Y}^n] \\ &= \frac{[\boldsymbol{Y}^n \mid \boldsymbol{\beta}, \sigma_Z^2][\boldsymbol{\beta} \mid \sigma_Z^2][\sigma_Z^2]}{[\boldsymbol{\beta} \mid \boldsymbol{Y}^n, \sigma_Z^2]} . \end{aligned}$$

Each term in the numerator is known and the denominator has been previously deduced from (4.3.4). □

5

Space-Filling Designs for Computer Experiments

5.1 Introduction

In this chapter and the next, we discuss how to *select inputs* at which to compute the output of a computer experiment to achieve specific goals. The inputs we select constitute our "experimental design." The region corresponding to the values of the inputs over which we wish to study or model the response is the experimental region. A point in this region corresponds to a specific set of values of the inputs. Thus, an experimental design is a specification of points in the experimental region at which we wish to compute the response.

We begin by reviewing some of the basic principles of classical experimental design and then present an overview of some of the strategies that have been employed in computer experiments. For details concerning the criteria used by classical design and methods of design construction see, for example, the books by Atkinson and Donev (1992), Box and Draper (1987), Dean and Voss (1999), Pukelsheim (1993), Silvey (1980), and Wu and Hamada (2000).

5.1.1 Some Basic Principles of Experimental Design

Suppose that we observe a response and wish to study how that response varies as we change a set of inputs. In physical experiments, there are a number of issues that make this problematic. First, the response may be affected by factors other than the inputs we have chosen to study. Unless we can completely control the effects of these additional factors, repeated observations at the same values of the inputs will vary as these additional factors vary. The effects of additional factors can either be unsystematic (random) or systematic. Unsystematic effects are usually referred to as random error, measurement error, or noise. Systematic effects are often referred to as bias. There are strategies for dealing with both noise and bias.

Replication and *blocking* are two techniques used to estimate and control the magnitude of random error. Replication (observing the response multiple times at the same set of inputs) allows one to directly estimate the magnitude and distribution of random error. Also, the sample means of replicated responses have smaller variances than the individual responses. Thus, the relation between these means and the inputs gives a clearer picture of the effects of the inputs because uncertainty from random error is reduced. In general, the more observations we have, the more information we have about the relation between the response and the inputs.

Blocking involves sorting experimental material into, or running the experiment in, relatively homogeneous sets called blocks. The corresponding analysis explores the relation between the response and the inputs within blocks, and then combines the results across blocks. Because of the homogeneity within a block, random error is less within a block than between blocks and the effects of the inputs more easily seen. There is an enormous body of literature on block designs, including both statistical and combinatorial issues. General discussions include John (1980), John (1987), Raghavarao (1971), or Street and Street (1987).

Bias is typically controlled by *randomization* and by exploring how the response changes as the inputs change. Randomization is accomplished by using a well-defined chance mechanism to assign the input values as well as any other factors that may affect the response and that are under the control of the experimenter, such as the order of experimentation, to experimental material. Factors assigned at random to experimental material will not systematically affect the response. By basing inferences on changes in the response as the input changes, bias effects "cancel," at least on average. For example, if a factor has the same effect on every response, subtraction (looking at changes or differences) removes the effect.

Replication, blocking, and randomization are basic principles of experimental design for controlling noise and bias. However, noise and bias are not the only problems that face experimenters. Another problem occurs when we are interested in studying the effects of several inputs simultaneously and the inputs themselves are highly correlated. This sometimes occurs in observational studies. If, for example, the observed values of two inputs are positively correlated so that they simultaneously increase, then it is difficult to distinguish their effects on the response. Was it the increase in just one or some combination of both that produced the observed change in the response? This problem is sometimes referred to as collinearity. Orthogonal designs are used to overcome this problem. In an orthogonal design, the values of the inputs at which the response is observed are uncorrelated. An orthogonal design allows one to independently assess the effects of the different inputs. There is a large body of literature on finding orthogonal designs, generally in the context of factorial experiments. See, for example, Hedayat, Sloane and Stufken (1999).

Another problem that can be partly addressed (or at least detected) by careful choice of an experimental design, occurs when the assumptions we make about the nature of the relation between the response and the inputs (our statistical model) are incorrect. For example, suppose we assume that the relationship between the response and a single input is essentially linear when, in fact, it is highly nonlinear. Inferences based on the assumption that the relationship is linear will be incorrect. It is important to be able to detect strong nonlinearities and we need to observe the response for at least three different values of the input in order to do so. Error that arises because our assumed model is incorrect is sometimes referred to as model bias. Diagnostics, such as scatterplots and quantile plots, are used to detect model bias. The ability to detect model bias is improved by careful choice of an experimental design, for example, by observing the response at a wide variety of values of the inputs. We would like to select designs that enable us to detect model inadequacies and lead to inferences that are relatively insensitive to model bias. This usually requires specifying both the model we intend to fit to our data as well as the form of an alternative model whose bias we wish to guard against; thus designs for model bias are selected to protect against certain types of bias. Box and Draper (1987) discuss this issue in more detail.

In addition to general principles, such as replication, blocking, randomization, orthogonality, and the ability to detect model bias, there exist very formal approaches to selecting an experimental design. The underlying principle is to consider the purpose of the experiment and choose the design accordingly. If we can formulate the purpose of our experiment in terms of optimizing a particular quantity, we can then ask what inputs we should observe the response at to optimize this quantity. For example, if we are fitting a straight line to data, we might wish to select our design so as to give us the most precise (minimum variance) estimate of the slope. This approach to selection of an experimental design is often referred to as *optimal design*. See Atkinson and Donev (1992), Pukelsheim (1993), or Silvey (1980) for more on the theory of optimal design. In the context of the linear model, popular criteria involve minimizing some function of the covariance matrix of the least squares estimates of the parameters. Some common functions are the determinant of the covariance matrix (the generalized variance), the trace of the covariance matrix (the average variance), and the average of the variance of the predicted response over the experimental region. A design minimizing the first criterion is called *D-optimal*, a design minimizing the second is called *A-optimal*, and a design minimizing the third is called *I-optimal.* In many experiments, especially experiments with multiple objectives, it may not be clear how to formulate the experiment goal in terms of some quantity that can be optimized. Furthermore, even if we can formulate the problem in this way, finding the optimal design may be quite difficult.

5.1.2 Design Strategies for Computer Experiments

Computer experiments, at least as we consider them here, differ from traditional physical experiments in that repeated observations at the same set of inputs yield identical responses. A single observation at a given set of inputs gives us perfect information about the response at that set of inputs, so replication is unnecessary. Uncertainty arises in computer experiments because we do not know the exact functional form of the relationship between the inputs and the response, although the response can be computed at any given input. Any functional models that we use to describe the relationship are only approximations. The discrepancy between the actual response produced by the computer code and the response we predict from the model we fit is our error. In the previous subsection we referred to such error as model bias.

Based on these observations, two principles for selecting designs are the following.

1. *Designs should not take more than one observation at any set of inputs. (But note that this principle assumes the computer code remains unchanged over time. When a design is run sequentially and the computer code is written and executed by a third party, it may be good policy to duplicate one of the design points in order to verify that the code has not been changed over the course of the experiment.)*

2. *Because we don't know the true relation between the response and inputs, designs should allow one to fit a variety of models and should provide information about all portions of the experimental region.*

If we believe that interesting features of the true model are just as likely to be in one part of the experimental region as another, it is plausible to use designs that spread the points at which we observe the response evenly throughout the region. There are a number of ways to define what it means to spread points evenly throughout a region and these lead to various types of designs. We discuss a number of these in this chapter. Among the designs we will consider are designs based on selecting points in the experimental region by certain sampling methods; designs based on measures of distance between points that allow one to quantify how evenly spread out points are; designs based on measures of how close points are to being uniformly distributed throughout a region; and designs that are a hybrid of or variation on these designs. We will refer to all the designs in this chapter as *space-filling* or *exploratory* designs.

Although we don't know the true model that describes the relation between the inputs and the response, if the models we fit to the data come from a sufficiently broad class, we may be willing to assume some model in this class is (to good approximation) correct. In this case it is possible to formulate specific criteria for choosing a design and adopt an optimal design approach. Because the models considered in the previous chapters are remarkably flexible, this approach seems reasonable for these models. Thus, we discuss some criterion-based methods for selecting designs in Chapter 6.

5.2 Designs Based on Methods for Selecting Random Samples

In the language of Subsection 2.2.1, the designs described in this section are used in cases when all inputs $\boldsymbol{x}$ are control variables as well as in cases when they are mixtures of control and environmental variables. However, most of these designs were originally motivated by their usefulness in applications where all the inputs were environmental variables; in this case we denote the inputs by $\boldsymbol{X}$ to emphasize their random nature. Let $y(\cdot)$ denote the output of the code. In this case, the most comprehensive objective would be to find the distribution of the random variable $Y = y(\boldsymbol{X})$ when $\boldsymbol{X}$ has a known distribution. If, as is often the case, this is deemed too difficult, the easier problem of determining some aspect of its distribution, such as its mean or its variance is considered. Several of the designs introduced in this section, in particular the Latin hypercube design, were introduced to solve the problem of estimating the mean of Y in such a setting. However, the reader should bear in mind that such designs are useful in more general input settings.

5.2.1 Designs Generated by Elementary Methods for Selecting Samples

Intuitively, we would like designs for computer experiments to be space-filling when prediction accuracy is of primary interest. The reason for this is that interpolators are used as predictors (e.g., the BLUP (3.3.4) or its Bayesian counterparts such as those that arise as the means of the predictive distributions derived in Section 4.1). Hence, the prediction error at any input site is a function of its location relative to the design points. Indeed, we saw, in Section 4.1, that the prediction error is *zero* at each of the design points. For this reason, designs that are not space-filling, for example, designs that concentrate points on the boundary of the design space, can yield predictors that perform quite poorly in portions of the experimental region that are sparsely observed.

One deterministic strategy for selecting the values of the inputs at which to observe the response is to choose these values so they are spread evenly throughout ("fill") the experimental region. There are several methods that might be used to accomplish this, depending on what one means by "spreading points evenly" or "filling the experimental region."

A very simple strategy is to select points according to a regular grid pattern superimposed on the experimental region. For example, suppose the experimental region is the unit square $[0,1] \times [0,1]$. If we wish to observe the response at 25 evenly spaced points, we might consider the grid of points $\{0.1, 0.3, 0.5, 0.7, 0.9\} \times \{0.1, 0.3, 0.5, 0.7, 0.9\}$.

There are several statistical strategies that one might adopt to fill a given experimental region. One possibility is to select a *simple random sample* of points from the experimental region. In theory, there are infinitely many points between 0 and 1 and this makes selecting a simple random sample problematic. In practice, we only record numbers to a finite number of decimal places and thus, in practice, the number of points between 0 and 1 can be regarded as finite. Therefore, we can assume our experimental region consists of finitely many points and select a simple random sample of these.

Simple random sampling in computer experiments can be quite useful. If we sample the inputs according to some distribution (for example, a distribution describing how the values are distributed in a given population), we can get a sense of how the corresponding outputs are distributed and this can serve as the basis for inferences about the distribution of the output. However, for many purposes, other sampling schemes, such as stratified random sampling, are preferable to simple random sampling. Even if the goal is simply to guarantee that the inputs are evenly distributed over the experimental region, simple random sampling is not completely satisfactory, especially when the sample sizes are relatively small. With small samples in high-dimensional experimental regions, the sample will typically exhibit some clustering and fail to provide points in large portions of the region.

In *stratified random sampling*, a set of n points is obtained by dividing the experimental region into n strata, spread evenly throughout the experimental region, and randomly selecting a single point from each. Varying the size and position of the strata, as well as sampling according to different distributions within the strata, allows considerable flexibility in selecting a design. This may be more or less useful, depending on the purpose of the computer experiment. For example, we may wish to explore some portions of the experimental region more thoroughly than others. However, if the goal is simply to select points that are spread evenly throughout the experimental region, spacing the strata evenly and sampling each according to a uniform distribution would seem the most natural choice.

If we expect the output to depend on only a few of the inputs (this is sometimes referred to as factor sparsity), then we might want to be sure that points are evenly spread across the projection of our experimental region onto these factors. A design that spreads points evenly throughout the full experimental region will not necessarily have this property. Alternatively, if we believe our model is well approximated by an additive model, a design that spreads points evenly across the range of each individual input (one-dimensional projection) might be desirable. For a sample of size n, it can be difficult to guarantee that a design has such projection properties, even with stratified sampling. Latin hypercube sampling, which we now discuss, is a way to generate designs that spread observations evenly over the range of each input separately.

5.2.2 Designs Generated by Latin Hypercube Sampling

Designs generated by Latin hypercube sampling are called Latin hypercube designs (LHD) throughout this book. We begin by introducing Latin hypercube (LH) sampling when the experimental region is the unit square $[0,1]^2$. To obtain a design consisting of n points, divide each axis $[0,1]$ into the n equally spaced intervals $[0,1/n), \ldots, [(n-1)/n, 1]$. This partitions the unit square into n^2 cells of equal size. Now, fill these cells with the integers 1, 2, ..., n so as to form a Latin square, i.e., an arrangement in which each integer appears exactly once in each row and in each column of this grid of cells. Select one of the integers at random. In each of the n cells containing this integer, select a point at random. The resulting sample of n points are a LHD of size n (see Figure 5.2 for an example with $n = 5$). The method of choosing the sample ensures that points are (marginally) spread evenly over the values of each input variable. Of course, such a LH sample could select points that are spread evenly along the diagonal of the square (see Figure 5.3). Although the points in such a sample have projections that are evenly spread out over the values of each input variable separately, we would not regard them as evenly spread out over the entire unit square.

We now describe a general procedure for obtaining a LH sample of size n from $\boldsymbol{X} = (X_1, \ldots, X_d)$ when $\boldsymbol{X}$ has independently distributed components. Stein (1987) discusses the implementation of LH sampling when $\boldsymbol{X}$ has dependent components, but we will not consider the dependent case here.

In the independence case the idea is as follows. Suppose that a LH sample of size n is to be selected. The domain of each input variable is divided into n intervals. Each interval will be represented in the LH sample. The set of all possible Cartesian products of these intervals constitutes a partitioning of the d-dimensional sample space into n^d "cells." A set of n cells is chosen from the n^d population of cells in such a way that the projections of the centers of each of the cells onto each axis are uniformly spread across the axis; then a point is chosen at random in each selected cell.

In detail, we construct the LH sample as follows. For $k = 1, \ldots, d$, let $F_k(\cdot)$ denote the (marginal) distribution of X_k, the kth component of $\boldsymbol{X}$ and, for simplicity, assume that X_k has support $[a_k, b_k]$. We divide the kth axis into n parts, each of which has equal probability, $1/n$, under $F_k(\cdot)$. The division points for the kth axis are

$$F_k^{-1}\left(\frac{1}{n}\right), \ldots, F_k^{-1}\left(\frac{n-1}{n}\right).$$

To choose n of the cells so created, let $\boldsymbol{\Pi} = (\Pi_{jk})$ be an $n \times d$ matrix having columns which are d different randomly selected permutations of $\{1, 2, \ldots, n\}$. Then the "upper-left hand" coordinates of the jth cell in $\mathbb{R}^d$ are

$$F_k^{-1}(n^{-1}(\Pi_{jk} - 1)), \quad k = 1, \ldots, d,$$

with the convention $F_k^{-1}(0) = a_k$.

For $j = 1, \ldots, n$, let X_{jk}, $k = 1, \ldots, d$, denote the kth component of the jth vector, $\boldsymbol{X}_j$. Then we define the LH sample to have values

$$X_{jk} = F_k^{-1}\left(\frac{1}{n}(\Pi_{jk} - 1 + U_{jk})\right),$$

where the $\{U_{jk}\}$ are independent and identically distributed $U[0, 1]$ deviates, for $j = 1, \ldots, n$ and $k = 1, \ldots, d$. In sum, the jth row of $\boldsymbol{\Pi}$ identifies the cell that $\boldsymbol{X}_j$ is sampled from, while the corresponding (independently generated) uniform deviates determine the location of $\boldsymbol{X}_j$ within the sampled cell.

Example 5.1 Suppose $\boldsymbol{X} = (X_1, X_2)$ is uniformly distributed over $[0, 1] \times [0, 1]$ so that $F_k^{-1}(w) = w, 0 < w < 1$. To obtain a LH sample of size $n = 3$, we compute

$$X_{jk} = \frac{1}{3}(\Pi_{jk} - 1 + U_{jk}), \quad j = 1, 2, 3;\ k = 1, 2.$$

To envision the pattern of a LH sample, divide the unit interval in each dimension into [0,1/3), [1/3,2/3), and [2/3,1], yielding a partition of $[0, 1] \times [0, 1]$ into nine squares (cells) of equal area. In the LH sample, each of these subintervals will be represented exactly once *in each dimension*. For simplicity of discussion, suppose we label these subintervals as 1, 2, and 3 in the order given above. One possible LHD would involve points randomly sampled from the (1,1), (2,3), and (3,2) squares and another possible design from the (1,2), (2,3), and (3,1) squares. Figure 5.1 plots the cells selected by the second design. These two selections correspond to the permutations

$$\boldsymbol{\Pi} = \begin{pmatrix} 1 & 1 \\ 2 & 3 \\ 3 & 2 \end{pmatrix} \text{ and } \boldsymbol{\Pi} = \begin{pmatrix} 1 & 2 \\ 2 & 3 \\ 3 & 1 \end{pmatrix}. \tag{5.2.1}$$

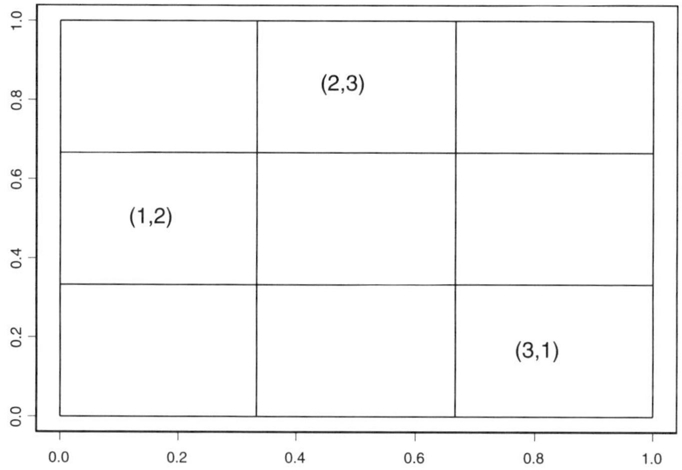

FIGURE 5.1. Cells selected by the Latin hypercube sample (1,2), (2,3), and (3,1)

Note that in each dimension, each subinterval appears exactly once. Because each subinterval is of length 1/3, the addition of $U_{jk}/3$ to the left-hand boundary of the selected subinterval serves merely to pick a specific point in it. ■

In the computer experiment setting, the input variables $\boldsymbol{x} = (x_1, x_2, \ldots, x_d)$ are not regarded as random for purposes of experimental design. As in Example 5.1, suppose that each input variable has been scaled to have domain [0,1]. Denoting the kth component of $\boldsymbol{x}_j$ by x_{jk} for $k = 1, \ldots, d$, suppose that we obtain an LHD from a given $\boldsymbol{\Pi}$ as follows:

$$x_{jk} = \frac{\Pi_{jk} - 0.5}{n}, \quad j = 1, \ldots, n;\ k = 1, \ldots, d.$$

This corresponds to taking $U_{jk} = 0.5$ for each $j = 1, \ldots, n$ and $k = 1, \ldots, d$ rather than as a sample from a $U[0,1]$ distribution. The "cells" are now identified with all d-dimensional Cartesian products of the intervals $\{(0, 1/n], (1/n, 2/n], \ldots, ((n-1)/n, 1]\}$, and each $\boldsymbol{x}_j$ is sampled from the *center* of the cell indicated by the jth row of $\boldsymbol{\Pi}$. An example of an LHD for $n = 5$ and $d = 2$ is given in Figure 5.2 with its associated $\boldsymbol{\Pi}$ matrix.

As mentioned previously, LHDs need not be space-filling over the full experimental region. To illustrate this point, consider the LHD for $n = 5$ and $d = 2$ that is shown in Figure 5.3, which one might *not* view as space-filling. One consequence of computing responses at this set of inputs is that we would expect a predictor fitted using this design to generally perform

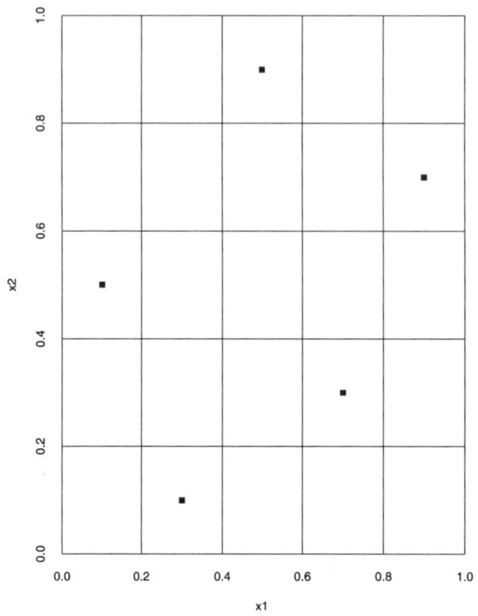

$$\Pi = \begin{pmatrix} 3 & 5 \\ 4 & 2 \\ 2 & 1 \\ 1 & 3 \\ 5 & 4 \end{pmatrix}$$

FIGURE 5.2. A space-filling Latin hypercube design

well only for $x_1 \approx x_2$. For example, consider the deterministic function

$$y(x_1, x_2) = \frac{x_1}{1 + x_2}, \qquad \mathcal{X} = [0,1] \times [0,1].$$

The MLE-EBLUP was fitted to the observed responses using the training data for both of the designs shown in Figures 5.2 and 5.3 (Subsection 3.3.2). The predictor was based on the stochastic process

$$Y(x_1, x_2) = \beta_0 + Z(x_1, x_2),$$

where $Z(\cdot)$ is a zero mean Gaussian stochastic process with unknown process variance and product power exponential correlation function (2.3.14). The prediction error $|y(x_1, x_2) - \widehat{Y}(x_1, x_2)|$ was calculated on a grid of 100 equally-spaced (x_1, x_2) points for each design. Figure 5.4 plots a comparison of the prediction errors for the two designs where the symbol "1" ("0") indicates that the prediction error for the design of Figure 5.3 is larger (smaller) than the prediction error for the design of Figure 5.2. The space-filling design of Figure 5.2 clearly yields a better predictor over most of the design space except for the diagonal where the LHD in Figure 5.3 collects most of its data.

It is apparent from Figure 5.3 and the subsequent discussion, that although all LHDs possess desirable *marginal* properties, only a subset of these designs are truly "space-filling." Subsection 5.5 will discuss design criteria that Welch (1985) has successfully applied to select space-filling LHDs for use in computer experiments.

are continuous analogs of an "analysis of variance" decomposition of $y(\boldsymbol{x})$. An additional reason for calling this an ANOVA decomposition is that

$$\int_{a_j}^{b_j} \alpha_j(x_j)\, dF_j(x_j) = 0 \quad \text{and} \quad \int_{\mathcal{X}_{-j}} r(\boldsymbol{x})\, dF_{-j}(\boldsymbol{x}_{-j}) = 0$$

for all j.

Stein (1987) shows that for large samples, the variance of $\overline{Y}$ is smaller under LH sampling than simple random sampling unless all main effect functions are 0. To be precise, let $\text{Var}_{LHS}\left\{\overline{Y}\right\}$ and $\text{Var}_{SRS}\left\{\overline{Y}\right\}$ denote the variance of $\overline{Y}$ under LH and simple random sampling, respectively. Stein (1987) proves the following expansions for the variance of $\overline{Y}$ under the two sampling schemes.

Theorem 5.2.2 As $n \to \infty$, under Latin hypercube sampling and simple random sampling we have

$$\begin{aligned}
\text{Var}_{LHS}\left\{\overline{Y}\right\} &= \frac{1}{n}\int_{\mathcal{X}} r^2(\boldsymbol{x}) dF(\boldsymbol{x}) + o(n^{-1}) \quad \text{and} \\
\text{Var}_{SRS}\left\{\overline{Y}\right\} &= \frac{1}{n}\int_{\mathcal{X}} r^2(\boldsymbol{x}) dF(\boldsymbol{x}) + \frac{1}{n}\sum_{i=1}^{d}\int_{a_i}^{b_i} \alpha_i^2(x_i) dF_i(x_i) + o(n^{-1}),
\end{aligned}$$

respectively.

The implication of this expansion is that, unless all $\alpha_j(\cdot)$ are identically 0, in the limit, LH sampling has a smaller variance than simple random sampling.

Further, not only can the variance of Y be estimated but normality of $\overline{Y}$ can be established. For simplicity, we assume $\mathcal{X} = [0,1]^d$ and that $F(\cdot)$ is uniform. More general cases can often be reduced to this setting by appropriate transformations. Owen (1992a) shows that $\overline{Y}$ computed from inputs based on LH sampling is approximately normally distributed for large samples. This can be used as the basis for statistical inference about μ. Owen (1992a) proves the following.

Theorem 5.2.3 If $y(\boldsymbol{x})$ is bounded then under LH sampling, $\sqrt{n}(\overline{Y} - \mu)$ tends in distribution to $N\left(0, \int_{\mathcal{X}} r^2(\boldsymbol{x}) d\boldsymbol{x}\right)$ as $n \to \infty$.

Owen (1992a) also provides estimators of the asymptotic variance

$$\int_{\mathcal{X}} r^2(\boldsymbol{x})\, d\boldsymbol{x}$$

to facilitate application of these results to computer experiments.

Subsection 5.6.3 of the Chapter Notes describes the use of LHDs in a generalization of these constant mean results to a regression setting, which has potential for use in computer experiments.

5.2.4 Variations on Latin Hypercube Designs

There are several ways in which LHDs have been extended. Randomized orthogonal arrays are one such extension. An *orthogonal array O on s symbols of strength t* is an $n \times p$ ($p \geq t$) matrix whose entries are the s symbols arranged so that in every $n \times t$ submatrix of O, all of the s^t possible rows appear the same number, λ, of times; obviously $n = \lambda s^t$. For additional discussion regarding orthogonal arrays see Raghavarao (1971) or Wu and Hamada (2000).

Owen (1992b) describes a procedure for generating n point space-filling designs in p dimensions from the columns of an $n \times p$ orthogonal array. The resulting designs are called *randomized orthogonal arrays.* If one plots the points of a randomized orthogonal array generated from an orthogonal array of strength t, in t or fewer of the coordinates, the result will be a regular grid. Thus these designs have desirable projection properties in higher dimensions that LHDs possess in each single dimension.

Example 5.2 A simple example of a randomized orthogonal array is the following. Suppose we take $n = 3$, $p = 2$, $s = 3$, $t = 1$, and $\lambda = 1$. An orthogonal array on three symbols of strength $t = 1$ is the 3×2 matrix, both of whose columns are the integers 1, 2, 3.

$$\begin{pmatrix} 1 & 1 \\ 2 & 2 \\ 3 & 3 \end{pmatrix}$$

From this, a randomized orthogonal array is generated by following the procedure described in Example 5.1 used to generate the design displayed in Figure 5.1 (permute the second column of the above orthogonal array using the second permutation in (5.2.1)). Because $t = 1$, the resulting design is an LHD and the projections onto each of the two dimensions (inputs) are uniform. Notice that, in general, an orthogonal array on s symbols of strength $t = 1$ with $n = s$ and $\lambda = 1$ is the $n \times p$ matrix, all of whose columns are the integers $1, 2, \ldots, s$. By following the procedure described in Subsection 5.2.2, one can generate a randomized orthogonal array which, in fact, is an LHD in p dimensions. ■

Example 5.3 Another example of a randomized orthogonal array is the following. Suppose we take $n = 9$, $p = 3$, $s = 3$, $t = 2$, and $\lambda = 1$. An orthogonal array on three symbols of strength $t = 2$ is the 9×3 matrix

$$\begin{pmatrix} 1 & 1 & 1 \\ 1 & 2 & 2 \\ 1 & 3 & 3 \\ 2 & 1 & 2 \\ 2 & 2 & 3 \\ 2 & 3 & 1 \\ 3 & 1 & 3 \\ 3 & 2 & 1 \\ 3 & 3 & 2 \end{pmatrix}$$

To construct a randomized orthogonal array, we use this 9×3 matrix. Divide the unit cube $[0,1] \times [0,1] \times [0,1]$ into a $3 \times 3 \times 3$ grid of 27 cells (cubes). Let (1,1,1) denote the cell (cube) $[0,1/3] \times [0,1/3] \times [0,1/3]$, (1,1,2) denote the cell $[0,1/3] \times [0,1/3] \times [1/3,2/3]$, (1,1,3) denote the cell $[0,1/3] \times [0,1/3] \times [2/3,1]$, ..., and (3,3,3) the cell $[2/3,1] \times [2/3,1] \times [2/3,1]$. Each row of the above 9×3 matrix corresponds to one of these 27 cells. The point in the center of the nine cells determined by the rows of the matrix yields a nine point randomized orthogonal array. Projected onto each two-dimensional subspace, the design looks like a regular 3×3 grid. Instead of selecting the points in the centers of the nine cells, one could select a point a random from each of these cells. The resulting projections onto two-dimensional subspaces would not be a regular grid, but would be evenly spaced in each of the two-dimensional subspaces. ■

Although randomized orthogonal arrays extend the projection properties of LHDs to more than one dimension, they have the drawback that they only exist for certain values of n, namely for $n = \lambda s^t$, and for certain values of p. Also, because $n = \lambda s^t$, only for relatively small values of s and t will the designs be practical for use in computer experiments in which individual observations are time-consuming to obtain and hence for which n must be small. See Tang (1993) and Tang (1994) for additional information about the use of randomized orthogonal arrays in computer experiments.

Cascading LHDs are another extension of LHDs. Cascading LHDs are introduced in Handcock (1991) and can be described as follows. Generate an LHD. At each point of this design, consider a small region around the point. In this small region, generate a second LHD. The result is a cluster of small LHDs and is called a *cascading Latin hypercube design*. Such designs allow one to explore both the local (in small subregions) and the global (over the entire experimental region) correlation structure of the model for the response.

5.3 Designs Based on Measures of Distance

In this subsection, we consider criteria for selecting a design that are based on a measure or metric that quantifies how spread out a set of points are. One way in which the points in a design might be considered to be spread out is for no two points in the design to be "too" close together. To be more precise, let $\mathcal{D} \subset \mathcal{X} \subset \mathrm{IR}^d$ be an arbitrary n-point design consisting of *distinct* input sites $\{\boldsymbol{x}_1, \boldsymbol{x}_2, \ldots, \boldsymbol{x}_n\}$. Let ρ be a metric on $\mathcal{X}$. For example, one important distance measure is p^{th} order distance between $\boldsymbol{w}, \boldsymbol{x} \in \mathcal{X}$ which is defined, for $p \geq 1$, by

$$\rho_p(\boldsymbol{w}, \boldsymbol{x}) = \left[\sum_{j=1}^{d} |w_j - x_j|^p \right]^{1/p} . \qquad (5.3.1)$$

Rectangular and Euclidean distance are $\rho_p(\cdot, \cdot)$ for $p = 1$ and $p = 2$, respectively.

One measure of the closeness of the points in the set $\mathcal{D}$ is the smallest distance between any two points in $\mathcal{D}$, i.e.,

$$\min_{\boldsymbol{x}_1, \boldsymbol{x}_2 \in \mathcal{D}} \rho_p(\boldsymbol{x}_1, \boldsymbol{x}_2).$$

A design that maximizes this measure is said to be a *maximin distance design* and denoted by $\mathcal{D}_{Mm}$; thus

$$\min_{\boldsymbol{x}_1, \boldsymbol{x}_2 \in \mathcal{D}_{Mm}} \rho_p(\boldsymbol{x}_1, \boldsymbol{x}_2) = \max_{\mathcal{D} \subset \mathcal{X}} \min_{\boldsymbol{x}_1, \boldsymbol{x}_2 \in \mathcal{D}} \rho_p(\boldsymbol{x}_1, \boldsymbol{x}_2).$$

In an intuitive sense, $\mathcal{D}_{Mm}$ designs guarantee that no two points in the design are "too close," and hence the design points are spread over $\mathcal{X}$.

A second way in which the points in the design $\mathcal{D}$ might be regarded as spread out over $\mathcal{X}$ is for *every* point in the space $\mathcal{X}$ to be "close" to *some* point in $\mathcal{D}$. To make this precise, define the distance between an arbitrary input site $\boldsymbol{x} \in \mathcal{X}$ and the design $\mathcal{D} \subset \mathcal{X}$ by

$$\rho_p(\boldsymbol{x}, \mathcal{D}) = \min_{\boldsymbol{x}_i \in \mathcal{D}} \rho_p(\boldsymbol{x}, \boldsymbol{x}_i) .$$

An n-point design $\mathcal{D}_{mM}$ is defined to be a *minimax distance design* if the maximum distance between arbitrary points $\boldsymbol{x} \in \mathcal{X}$ and the candidate design $\mathcal{D}_{mM}$ is a minimum over all designs $\mathcal{D} \subset \mathcal{X}$, namely

$$\min_{\mathcal{D} \subset \mathcal{X}} \max_{\boldsymbol{x} \in \mathcal{X}} \rho_p(\boldsymbol{x}, \mathcal{D}) = \max_{\boldsymbol{x} \in \mathcal{X}} \rho_p(\boldsymbol{x}, \mathcal{D}_{mM}).$$

Another more general approach to the problem of spreading points out in the design space is to consider criteria that minimize the "average" of some function of the distances between pairs of design points. To simplify the description below, normalize the domain of each input variable to the interval [0,1] so that $\mathcal{X} = [0,1]^d$. With this convention, the distance calculations will be comparable across dimensions. Let $\mathcal{D} \subset \mathcal{X}$ be an arbitrary n-point design consisting of distinct input sites $\{\boldsymbol{x}_1, \boldsymbol{x}_2, \ldots, \boldsymbol{x}_n\}$. We define the average distance criterion function based on $\rho_p(\cdot,\cdot)$ as follows,

$$m_{(p,\lambda)}(\mathcal{D}) = \left(\frac{1}{\binom{n}{2}} \sum_{\boldsymbol{x}_i, \boldsymbol{x}_j \in D} \left[\frac{d^{1/p}}{\rho_p(\boldsymbol{x}_i, \boldsymbol{x}_j)} \right]^\lambda \right)^{1/\lambda}, \quad \lambda \geq 1. \tag{5.3.2}$$

The combinatorial coefficient $\binom{n}{2} = n(n-1)/2$ is the number of different pairs of points $(\boldsymbol{x}_i, \boldsymbol{x}_j)$ that can be drawn from the n points in $\mathcal{D}$. In general, for integer j with $0 \leq j \leq n$, $\binom{n}{j} = n!/(j!(n-j)!)$. For example, when $\lambda = 1$, the criterion function $m_{(p,1)}(\mathcal{D})$ is inversely proportional to the harmonic mean of the distances between all pairs of design points.

The value $d^{1/p}$ in the numerator of (5.3.2) is the maximum $\rho_p(\cdot,\cdot)$ distance between points in $[0,1]^d$, i.e.,

$$0 < \rho_p(\boldsymbol{x}_1, \boldsymbol{x}_2) \leq d^{1/p}$$

for all $\boldsymbol{x}_1 \neq \boldsymbol{x}_2 \in [0,1]^d$. This normalization makes distances comparable for pairs of points in different dimensions.

For fixed (p, λ), an n-point design $\mathcal{D}_{av}$ is optimal with respect to criterion (5.3.2) if

$$m_{(p,\lambda)}(\mathcal{D}_{av}) = \min_{\mathcal{D} \subset \mathcal{X}} m_{(p,\lambda)}(\mathcal{D}). \tag{5.3.3}$$

This optimality condition favors designs that possess nonredundancy in the location of input sites. For example, when $\lambda = 1$, the optimality condition (5.3.3) selects designs that maximize the harmonic mean, preventing any "clumping" of design points.

The nonredundancy requirement can be seen even more clearly for large values of λ. Taking $\lambda \to \infty$, the criterion function (5.3.2) becomes

$$m_{(p,\infty)}(\mathcal{D}) = \max_{\boldsymbol{x}_i, \boldsymbol{x}_j \in \mathcal{D}} \frac{d^{1/p}}{\rho_p(\boldsymbol{x}_i, \boldsymbol{x}_j)}.$$

An n-point design $\mathcal{D}_{Mm}$ will be a maximin distance design if condition (5.3.3) is satisfied for $\lambda = \infty$,

$$m_{(p,\infty)}(\mathcal{D}_{Mm}) = \min_{\mathcal{D} \subset \mathcal{X}} m_{(p,\infty)}(\mathcal{D}),$$

because this criterion is equivalent to maximizing the minimum distance between all pairs of design points,

$$\max_{\mathcal{D} \subset \mathcal{X}} \min_{\boldsymbol{x}_i, \boldsymbol{x}_j \in \mathcal{D}} \rho_p(\boldsymbol{x}_i, \boldsymbol{x}_j) \propto \frac{1}{m_{(p,\infty)}(\mathcal{D}_{Mm})}.$$

Johnson, Moore and Ylvisaker (1990) first defined the maximin distance design criterion.

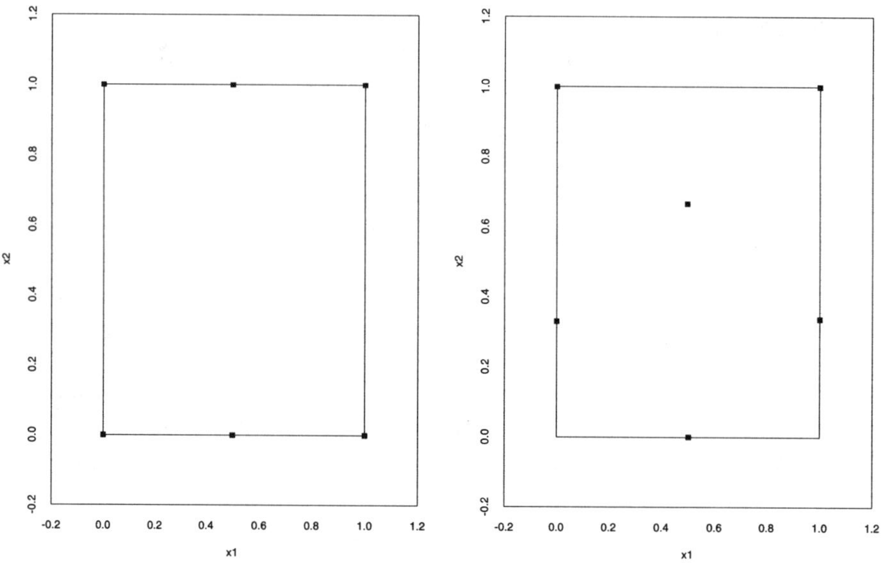

FIGURE 5.5. Optimal average Euclidean distance designs ($p = 2$) for $\lambda = 1.0$ (left panel) and $\lambda = \infty$ (right panel)

Example 5.4 Figure 5.5 displays optimal Euclidean distance designs for $\lambda = 1$ and $\lambda = \infty$ assuming $n = 6$ and $d = 2$. These designs concentrate points on or near the boundary of $\mathcal{X}$. We can remedy this by restricting the class of available designs to only include, say, LHDs. This provides a computationally convenient method of generating space-filling designs for computer experiments. Figure 5.2 is an example of an LHD that is optimal with respect to the average distance criterion (5.3.3) for $p = 1$ and $\lambda = 1$. The use of multiple criteria to select designs is discussed further in Subsection 5.2.4. ■

The optimal average distance designs described above need not have projections that are nonredundant. Consider a computer experiment involving $d = 5$ input variables, only three of which (say) are active. In this event, a desirable property of an optimal design is nonredundancy of input sites projected into three-dimensional subspaces of the full design space. Such designs can be generated by computing the criterion values (5.3.2) for each relevant projection of the full design $\mathcal{D}$ and averaging these to form a new criterion function which is then minimized by choice of design $\mathcal{D}$. This approach is implemented by the Algorithms for the Construction of Experimental Designs (ACED) software of Welch (1985).

Formally, let J denote the index set of subspace *dimensions* in which nonredundancy of input sites is desired. For example, $J = \{2, 3\}$ means we desire nonredundancy in all the 2 and 3 dimensional projections of the design points. For each $j \in J$, let $\{\mathcal{D}_{kj}\}$ denote the kth design in an enumeration of all j-dimensional projections of the n points in $\mathcal{D}$ for $k = 1, \ldots, \binom{d}{j}$. The elements of each $\{\mathcal{D}_{kj}\}$ are the projections of the $\boldsymbol{x} \in \mathcal{D}$ onto the j coordinates of the kth projection space. The average distance criterion function for the projected design $\mathcal{D}_{kj}$ is, from (5.3.2),

$$m_{(p,\lambda)}(\mathcal{D}_{kj}) = \left(\frac{1}{\binom{n}{2}} \sum_{\boldsymbol{x}_h^*, \boldsymbol{x}_i^* \in \mathcal{D}_{kj}} \left[\frac{j^{1/p}}{\rho_p(\boldsymbol{x}_h^*, \boldsymbol{x}_i^*)} \right]^{\lambda} \right)^{1/\lambda}, \tag{5.3.4}$$

for $k = 1, \ldots, \binom{d}{j}$ and $j \in J$. Here, $\boldsymbol{x}_\ell^*$ denotes the projection of $\boldsymbol{x}_\ell$ onto $\mathcal{D}_{kj}$.

As noted above, the normalization in the numerator of $m_{(p,\lambda)}(\cdot)$ ensures that the criterion values in (5.3.4) can be compared across subspaces of any dimension. With this in mind, we define the average projection design criterion function to be

$$av_{(p,\lambda)}(\mathcal{D}) = \left(\frac{1}{\sum_{j \in J} \binom{d}{j}} \sum_{j \in J} \sum_{k=1}^{\binom{d}{j}} [m_{(p,\lambda)}(\mathcal{D}_{kj})]^{\lambda} \right)^{1/\lambda}. \tag{5.3.5}$$

This function is an average of the criterion functions (5.3.4). An n-point design $\mathcal{D}_{avp}$ is average projection distance optimal with respect to criterion (5.3.5) if

$$\text{av}_{(p,\lambda)}(\mathcal{D}_{avp}) = \min_{\mathcal{D} \subset \mathcal{X}} \text{av}_{(p,\lambda)}(\mathcal{D}). \tag{5.3.6}$$

Example 5.5 The optimal average projection designs (5.3.6) will also be space-filling if the class of available designs is restricted to LHDs. As an example, let $n = 10$ and $d = 3$. An optimal average-projection LHD was generated with the specifications $p = \lambda = 1$ and $J = \{2, 3\}$. Figure 5.6 presents the full optimal design and its projection into the (x_2, x_3) subspace. Note that $1 \notin J$, as LHDs are nonredundant in each one-dimensional subspace by definition. ■

Notice that average projection designs are an alternative to LHDs and randomized orthogonal arrays for producing designs that are space-filling and are (reasonably) uniformly spread out when projected onto certain lower dimensional subspaces. Unlike randomized orthogonal arrays that only exist for certain values of n, average projection designs can be generated for any sample size.

5.4 Uniform Designs

In Section 5.2 we considered criteria for selecting a space-filling design based on sampling methods and, in Section 5.3, criteria based on distances

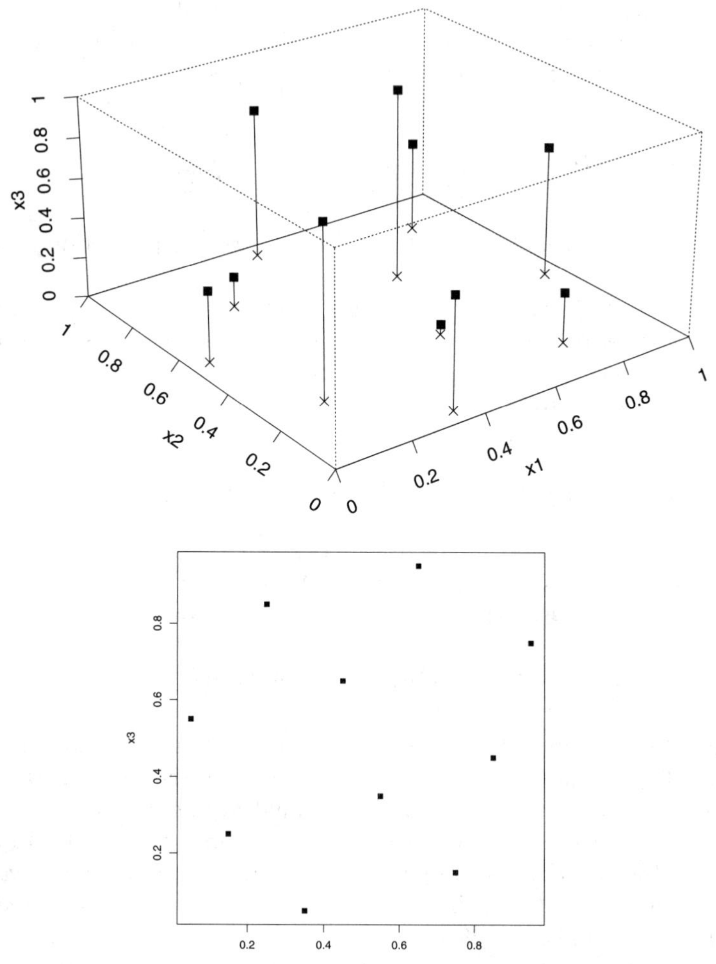

FIGURE 5.6. Top panel: an $n = 10$ point optimal average-projection Latin hypercube design in three dimensions when $p = \lambda = 1$ and $J = \{2, 3\}$. Bottom panel: projection of top panel onto x_2-x_3 plane.

between points. In this section, we consider a third intuitive design principle based on comparing the distribution of the points in a design to the uniform distribution.

As in Subsection 5.2.3, suppose that the vector of inputs is d-dimensional and denoted by $\boldsymbol{x} = (x_1, \ldots, x_d)$. Also again assume that $\boldsymbol{x}$ must fall in the d-dimensional hyper-rectangle $\mathcal{X} = \times_{i=1}^{d}[a_i, b_i]$. Let $\mathcal{D} = \{\boldsymbol{x}_1, \boldsymbol{x}_2, \ldots, \boldsymbol{x}_n\}$ denote the set of n points at which we will observe the response $y(\boldsymbol{x})$. If we wish to emphasize that $\boldsymbol{x}$ is a random variable, we will use the notation $\boldsymbol{X}$. This would be the case, for example, if we are interested in $\mathrm{E}\{y(\boldsymbol{X})\}$. Below we take $\boldsymbol{X} \sim F(\cdot)$ where

$$F(\boldsymbol{x}) = \prod_{i=1}^{d} \left(\frac{x_i - a_i}{b_i - a_i} \right) \tag{5.4.1}$$

is the uniform distribution on $\mathcal{X}$ (other choices of distribution function are possible).

Fang, Lin, Winker and Zhang (2000) discuss the notion of the *discrepancy* of a design $\mathcal{D}$, which measures the extent to which $\mathcal{D}$ differs from a completely uniform distribution of points. To be specific, let F_n be the empirical distribution function of the points in $\mathcal{D}$, namely

$$F_n(\boldsymbol{x}) = \frac{1}{n} \sum_{i=1}^{n} I\left\{\boldsymbol{X}_i \le \boldsymbol{x}\right\}, \tag{5.4.2}$$

where $I\{E\}$ is the indicator function of the event E and the inequality is with respect to the componentwise ordering of vectors in $\mathbb{R}^d$. The L_∞ discrepancy, sometimes called *star discrepancy* or simply discrepancy, is denoted $D_\infty(\mathcal{D})$ and is defined as

$$D_\infty(\mathcal{D}) = \sup_{\boldsymbol{x} \in \mathcal{X}} \mid F_n(\boldsymbol{x}) - F(\boldsymbol{x}) \mid . \tag{5.4.3}$$

This is perhaps the most popular measure of discrepancy and is the Kolmogorov-Smirnov statistic for testing fit to the uniform distribution.

Example 5.6 Suppose $d = 1$ and $\mathcal{X} = [0, 1]$ is the unit interval. It is not too difficult to show that the n point set

$$\mathcal{D} = \left\{ \frac{1}{2n}, \frac{3}{2n}, \ldots, \frac{2n-1}{2n} \right\}$$

has discrepancy $D_\infty(\mathcal{D}) = 1/2n$ because $F(x) = x$ in this case. ■

Another important measure of discrepancy is the L_p discrepancy of $\mathcal{D}$ which is denoted by $D_p(\mathcal{D})$ and defined by

$$D_p(\mathcal{D}) = \left[\int_{\mathcal{X}} |F_n(\boldsymbol{x}) - F(\boldsymbol{x})|^p \, d\boldsymbol{x} \right]^{1/p} . \tag{5.4.4}$$

The L_∞ discrepancy of $\mathcal{D}$ is a limiting case of L_p discrepancy obtained by letting p go to infinity.

Niederreiter (1992) discusses the use of discrepancy for generating uniformly distributed sequences of points in the context of quasi-Monte Carlo methods. Designs taking observations at sets of points with small discrepancies would be considered more uniform or more spread out than designs corresponding to sets with larger discrepancies. *Uniform designs* take observations at a set of points that minimizes D_p.

Other than the fact that it seems intuitively reasonable to use designs that are spread uniformly over $\mathcal{X}$, why might one consider using a uniform design? One reason that has been proposed is the following. Suppose we are interested in estimating the mean of $g(y(\boldsymbol{X}))$,

$$\mu = \mathrm{E}\{g(y(\boldsymbol{X}))\} = \int_{\mathcal{X}} g(y(\boldsymbol{x})) \frac{1}{\prod_{i=1}^{d}(b_i - a_i)} d\boldsymbol{x},$$

where $g(\cdot)$ is some known function. We consider the properties of the naíve moment estimator

$$T = T(y(\boldsymbol{X}_1), \ldots, y(\boldsymbol{X}_n)) = \frac{1}{n} \sum_{j=1}^{n} g(y(\boldsymbol{X}_j)).$$

The Koksma-Hlawka inequality (Niederreiter (1992)) gives an upper bound on the absolute error of this estimator, namely

$$| \, T(y(\boldsymbol{x}_1), \ldots, y(\boldsymbol{x}_n)) - \mu \, | \leq D_\infty(\mathcal{D}) V(g),$$

where $V(g)$ is a measure of the variation of g that does not depend on $\mathcal{D}$ (see page 19 of Niederreiter (1992) for the definition of $V(g)$). For fixed $g(\cdot)$, this bound is a minimum when $\mathcal{D}$ has minimum discrepancy. This suggests that a uniform design may control the maximum absolute error of T as an estimator of μ. Also, because this holds for any $g(\cdot)$, it suggests that uniform designs may be robust to the choice of $g(\cdot)$ because they have this property regardless of the value of $g(\cdot)$.

However, just because an upper bound on the absolute error is minimized, it does not necessarily follow that a uniform design minimizes the maximum absolute error over $\mathcal{X}$ or has other desirable properties. Furthermore, in the context of computer experiments, we are usually not interested in estimating μ. Thus, the above is not a completely compelling reason to use a uniform design in computer experiments as discussed here.

Wiens (1991) provides another reason for considering uniform designs. Suppose we believe the response $y(\boldsymbol{x})$ follows the regression model

$$y(\boldsymbol{x}) = \beta_0 + \sum_{i=1}^{k} \beta_i f_i(\boldsymbol{x}) + \varphi(\boldsymbol{x}) + \epsilon,$$

where the $\{f_i\}$ are known functions, the β_i unknown regression parameters, φ is an unknown function representing model bias, and ϵ normal random error. Wiens (1991) shows that under certain conditions on φ, the uniform design is best in the sense of maximizing the power of the overall F test of the regression.

Fang et al. (2000) provide yet another reason why one may wish to use uniform designs. They note that in orthogonal designs, the points are typically uniformly spread out over the design space. Thus, there is the possibility that uniform designs may often be orthogonal. To explore this further, they use computer algorithms to find designs that minimize a variety of measures of discrepancy and in doing so generate a number of orthogonal designs. Efficient algorithms for generating designs that minimize certain measures of discrepancy, therefore, may be useful in searching for orthogonal designs.

Fang et al. (2000) discuss a method for constructing (nearly) uniform designs. For simplicity, assume $\mathcal{X}$ is $[0,1]^d$. In general, finding a uniform design is not easy. One way to simplify the problem is to reduce the domain of $\mathcal{X}$. Obviously, a uniform design over this reduced domain may not be close to uniform over $\mathcal{X}$, but suitable selection of a reduced domain may yield designs which are nearly uniform. Based on the uniform design for $d = 1$, we might proceed as follows. Let $\mathbf{\Pi} = (\Pi_{ij})$ be an $n \times d$ matrix such that each column of $\mathbf{\Pi}$ is a permutation of the integers $\{1, 2, \ldots, n\}$. Let $\boldsymbol{X}(\mathbf{\Pi}) = (x_{ij})$ be the $n \times d$ matrix defined by

$$x_{ij} = (\Pi_{ij} - 0.5)/n$$

for all i, j. The n rows of $\boldsymbol{X}$ define n points in $\mathcal{X} = [0,1]^d$. Hence, each matrix $\mathbf{\Pi}$ determines an n point design. For example, when $d = 1$, if $\mathbf{\Pi} = (1, 2, \ldots, n)^\top$, then

$$\boldsymbol{X}(\mathbf{\Pi}) = \left(\frac{1}{2n}, \frac{3}{2n}, \ldots, \frac{2n-1}{2n}\right)^\top,$$

which is the uniform design in $d = 1$ dimension. Note that the n rows of $\boldsymbol{X}(\mathbf{\Pi})$ correspond to the sample points of an LHD with points at the centers of each sampled cell. One might search over the set $\mathcal{P}$ of all possible permutations $\mathbf{\Pi}$, selecting the $\mathbf{\Pi}$ that produces the n point design with minimum discrepancy. One would hope that this choice of design is nearly uniform over $\mathcal{X}$. Fang et al. (2000) describe two algorithms for conducting such a search (see Section 5 of their paper). Bratley, Fox and Niederreiter (1994) is an additional source for an algorithm that can be used to generate low-discrepancy sequences of points and hence (near) uniform designs.

The discrepancy D_∞ sometimes selects designs which intuitively do not appear to be uniform. The following example illustrates such a case.

Example 5.7 Suppose $d = 2$, $\mathcal{X} = [0,1]^2$, and consider the class of all designs generated by the set of permutations $\mathcal{P}$ introduced in the previous paragraph. One member of this class of designs is

$$\mathcal{D}_{diag} = \left\{ \left(\frac{1}{2n}, \frac{1}{2n}\right), \left(\frac{3}{2n}, \frac{3}{2n}\right), \ldots, \left(\frac{2n-1}{2n}, \frac{2n-1}{2n}\right) \right\}.$$

This n point design takes observations along the diagonal extending from the origin to the point $(1,1)$. Intuitively, we would expect $\mathcal{D}_{diag}$ to be a poor design, because it takes observations only along the diagonal and does not spread observations over $[0,1]^2$. To compute the discrepancy of $\mathcal{D}_{diag}$, we first compute the empirical distribution function F_n for $\mathcal{D}_{diag}$ at an arbitrary point $\boldsymbol{x} = (x_1, x_2)$ in $[0,1]^2$. Notice that points in $\mathcal{D}_{diag}$ have both coordinates equal and it is not too hard to show from Equation (5.4.2) that

$$F_n(x_1, x_2) = \frac{\text{number of pts. in } \mathcal{D}_{diag} \text{ with first coordinate} \leq \min\{x_1, x_2\}}{n}.$$

Notice that $F_n(\cdot,\cdot)$ is constant almost everywhere except for jumps of size $1/n$ at points for which one of the coordinates takes one of the values $\frac{1}{2n}, \frac{3}{2n}, \ldots, \frac{2n-1}{2n}$. In particular, $F_n(x_1, x_2)$ has value $\frac{m}{n}$ $(1 \leq m \leq n)$ on the set $\mathcal{X}_m$:

$$\left\{ (x_1, x_2) \in [0,1]^2 : \frac{2m-1}{2n} \leq \min\{x_1, x_2\} < \frac{2m+1}{2n} \right\}.$$

Recall from (5.4.1) that $F(\cdot)$ is the uniform distribution

$$F(\boldsymbol{x}) = x_1 x_2$$

on $\mathcal{X} = [0,1]^2$. On $\mathcal{X}_m$, the minimum value of $F(\boldsymbol{x})$ is $\left(\frac{2m-1}{2n}\right)^2$ and the supremum of $F(\boldsymbol{x})$ is $\frac{2m+1}{2n}$. This supremum is obtained in the limit as $\epsilon \to 0$ along the sequence of points $\left(\frac{2m+1}{2n} - \epsilon, 1\right)$. Thus, over $\mathcal{X}_m$, the supremum of $|F_n(\boldsymbol{x}) - F(\boldsymbol{x})|$ is either $\left|\frac{m}{n} - \left(\frac{2m-1}{2n}\right)^2\right|$ or $\left|\frac{m}{n} - \frac{2m+1}{2n}\right| = \frac{1}{2n}$. For $1 \leq m \leq n$, it is not difficult to show that

$$\left| \frac{m}{n} - \left(\frac{2m-1}{2n}\right)^2 \right| > \frac{1}{2n}.$$

Hence, over the set of all points $\boldsymbol{x}$ for which $F_n(\boldsymbol{x})$ has value $\frac{m}{n}$, the supremum of $|F_n(\boldsymbol{x}) - F(\boldsymbol{x})|$ is

$$\frac{m}{n} - \left(\frac{2m-1}{2n}\right)^2 = \frac{nm - m^2 + m}{n^2} - \frac{1}{4n^2},$$

continuous inputs while the MATLAB software assumes continuous inputs taking values in a hyper-rectanglar $\mathcal{X}$. Finally we mention the mathematical approach of Trosset (1999a) and Dimnaku, Kincaidy and Trosset (2002) who consider algorithms for constructing maximin distance designs for convex $\mathcal{X}$.

5.6 Chapter Notes

5.6.1 *Details of the Inequality (5.2.2) for Proportional Stratified Sampling with Random Sampling*

As in Subsection 5.2.3, suppose $\mathcal{X}_1, \mathcal{X}_2, \ldots, \mathcal{X}_I$ are disjoint subsets of $\mathcal{X}$ such that their union is $\mathcal{X}$; the $\mathcal{X}_i$ are the "strata" of $\mathcal{X}$. Let $p_i = P\{\boldsymbol{X} \subset \mathcal{X}_i\}$ denote the proportion of the sample space in the ith strata specified by the distribution $F(\cdot)$. Suppose that $F(\cdot)$ has density $f(\cdot)$.

Select a sample of size n_i from $\mathcal{X}_i$, , $i = 1, \ldots, I$, so that the sum of the n_i equals the total sample size n. This means that the sample from $\mathcal{X}_i$ has density given by the conditional density of $F(\cdot)$ on $\mathcal{X}_i$ which is

$$f(\boldsymbol{x} \mid \mathcal{X}_i) = \begin{cases} f(\boldsymbol{x})/p_i\,, & \text{if } \boldsymbol{x} \in \mathcal{X}_i \\ 0\,, & \text{otherwise} \end{cases}$$

and let $\boldsymbol{X}_{i1}, \boldsymbol{X}_{i2}, \ldots, \boldsymbol{X}_{in_i}$ denote this sample. We combine these separate samples to form the stratified estimator of μ,

$$T_S = \sum_{i=1}^{I} \left[\left(\frac{p_i}{n_i}\right) \sum_{j=1}^{n_i} g(y(\boldsymbol{X}_{ij})) \right].$$

To see that T_S is an unbiased estimator of μ, we let

$$\mu_i = \mathrm{E}\{g(y(\boldsymbol{X}_{ij}))\} = \int_{\mathcal{X}_i} g(y(\boldsymbol{x}))\frac{1}{p_i} f(\boldsymbol{x}) d\boldsymbol{x}$$

denote the conditional mean of an observation taken in the ith strata and

$$\sigma_i^2 = \mathrm{Var}\,\{g(y(\boldsymbol{X}_{ij}))\} = \int_{\mathcal{X}_i} [g(y(\boldsymbol{x})) - \mu_i]^2 \frac{1}{p_i} f(\boldsymbol{x}) d\boldsymbol{x}$$

the conditional variance of an observation taken in the ith strata. We write

$$\begin{aligned}
\mathrm{E}\{T_S\} &= \sum_{i=1}^{I}\left[\left(\frac{p_i}{n_i}\right)\sum_{j=1}^{n_i} E\{g(y(\boldsymbol{X}_{ij}))\}\right] \\
&= \sum_{i=1}^{I}\left[\left(\frac{p_i}{n_i}\right) n_i\mu_i\right] \\
&= \sum_{i=1}^{I} p_i\mu_i \\
&= \sum_{i=1}^{I}\left[p_i \int_{\mathcal{X}_i} g(y(\boldsymbol{x}))\frac{1}{p_i} f(\boldsymbol{x})d\boldsymbol{x}\right] \\
&= \int_{\mathcal{X}} g(y(\boldsymbol{x}))f(\boldsymbol{x})d\boldsymbol{x} = \mu.
\end{aligned}$$

Thus, T_S is an unbiased estimator of μ. Also, since samples from different strata are independent,

$$\mathrm{Var}\{T_S\} = \sum_{i=1}^{I}\left[\left(\frac{p_i^2}{n_i^2}\right)\sum_{j=1}^{n_i} \mathrm{Var}\{g(y(\boldsymbol{X}_{ij}))\}\right] = \sum_{i=1}^{I}\left(\frac{p_i^2}{n_i}\right)\sigma_i^2.$$

In the special case of proportional sampling,

$$T_S = \frac{1}{n}\sum_{i=1}^{I}\sum_{j=1}^{n_i} g(y(\boldsymbol{X}_{ij}))$$

which has the form of our general estimator T. In this case,

$$\begin{aligned}
\mathrm{Var}\{T_S\} &= \sum_{i=1}^{I}\left(\frac{p_i^2}{n_i}\right)\sigma_i^2 \\
&= \sum_{i=1}^{I}\left(\frac{p_i^2}{n_i}\right)\int_{\mathcal{X}_i} [g(y(\boldsymbol{x})) - \mu + \mu - \mu_i]^2 \frac{1}{p_i} f(\boldsymbol{x})d\boldsymbol{x} \\
&= \sum_{i=1}^{I}\left(\frac{p_i}{n_i}\right)\int_{\mathcal{X}_i} [(g(y(\boldsymbol{x})) - \mu)^2 \\
&\quad +2(g(y(\boldsymbol{x})) - \mu)(\mu - \mu_i) + (\mu - \mu_i)^2] f(\boldsymbol{x})d\boldsymbol{x} \\
&= \frac{1}{n}\left[\sum_{i=1}^{I}\int_{\mathcal{X}_i} (g(y(\boldsymbol{x})) - \mu)^2 f(\boldsymbol{x})d\boldsymbol{x} - \sum_{i=1}^{I} p_i(\mu - \mu_i)^2\right]
\end{aligned}$$

$$
\begin{aligned}
&= \frac{1}{n}\left[\int_{\mathcal{X}} (g(y(\boldsymbol{x})) - \mu)^2 f(\boldsymbol{x})d\boldsymbol{x} - \sum_{i=1}^{I} p_i(\mu - \mu_i)^2\right] \\
&= \text{Var}\{T_R\} - \frac{1}{n}\sum_{i=1}^{I} p_i(\mu - \mu_i)^2 \\
&\leq \text{Var}\{T_R\} \ . \ \square
\end{aligned}
$$

5.6.2 Proof That T_L is Unbiased and of Theorem 5.2.1

We use the same notation as in Subsection 5.2.3. To compute $\text{E}\{T_L\}$, we need to describe how the LH sample is constructed. For each i, $1 \leq i \leq d$, divide the range $[a_i, b_i]$ of the ith coordinate of $\boldsymbol{X}$ into n intervals of equal marginal probability $1/n$ under $F(\cdot)$. Sample once from each of these intervals and let these sample values be denoted $X_{i1}, X_{i2}, \ldots, X_{in}$. Form the $d \times n$ array

$$
\begin{pmatrix}
X_{11} & X_{12} & \ldots & X_{1n} \\
X_{21} & X_{22} & \ldots & X_{2n} \\
 & & \vdots & \\
X_{d1} & X_{d2} & \ldots & X_{dn}
\end{pmatrix}
$$

and then randomly permute the elements in each row using independent permutations. The n columns of the resulting array are the LH sample. This is essentially the procedure for selecting a LH sample that was discussed in Subsection 5.2.1.

Another way to select a LH sample is as follows. The Cartesian product of the d subintervals $[a_i, b_i]$ partitions $\mathcal{X}$ into n^d cells, each of probability $1/n^d$. Each of these n^d cells can be labeled by a set of d coordinates

$$
\boldsymbol{m}_i = (m_{i1}, m_{i2}, \ldots, m_{id}),
$$

where $1 \leq i \leq n^d$ and m_{ij} is a number between 1 and n corresponding to which of the n intervals of $[a_j, b_j]$ is represented in cell i. For example, suppose $n = 3$, $d = 2$, $[a_1, b_1] = [a_2, b_2] = [0, 1]$, and $F(\cdot)$ is uniform. We divide $[a_1, b_1]$ into the three intervals $[0, 1/3)$, $[1/3, 2/3)$, and $[2/3, 1]$. Similarly for $[a_2, b_2]$. In this case the cell $[1/3, 2/3) \times [1/3, 2/3)$ would have cell coordinates (2, 2). To obtain a LH sample, select a random sample of n of the n^d cells, say $\boldsymbol{m}_{i_1}, \boldsymbol{m}_{i_2}, \ldots, \boldsymbol{m}_{i_n}$, subject to the condition that for each j, the set $\{m_{i_\ell j}\}_{\ell=1}^{n}$ is a permutation of the integers $1, 2, \ldots, n$. We then randomly select a single point from each of these n cells.

For an LH sample obtained in this manner, the density of $\boldsymbol{X}$, given $\boldsymbol{X} \in$ cell i, is

$$
f(\boldsymbol{x} \mid \boldsymbol{X} \in \text{cell } i) = \begin{cases} n^d f(\boldsymbol{x}), & \text{if } \boldsymbol{x} \in \text{cell } i \\ 0, & \text{otherwise.} \end{cases}
$$

Thus, the distribution of the output $y(\boldsymbol{X})$ under LH sampling is

$$\begin{aligned} P(y(\boldsymbol{X}) \le y) &= \sum_{i=1}^{n^d} P(y(\boldsymbol{X}) \le y \mid \boldsymbol{X} \in \text{cell } i) P(\boldsymbol{X} \in \text{cell } i) \\ &= \sum_{i=1}^{n^d} \int_{\text{cell } i \text{ and } y(\boldsymbol{x}) \le y} n^d f(\boldsymbol{x}) \left(\frac{1}{n^d}\right) d\boldsymbol{x} \\ &= \int_{y(\boldsymbol{x}) \le y} f(\boldsymbol{x}) d\boldsymbol{x}, \end{aligned}$$

which is the same as for simple random sampling. Hence we have $\mathrm{E}\{T_L\} = \mu$.

To compute $\mathrm{Var}\{T_L\}$, we view our sampling as follows. First we select the $\boldsymbol{X}_i$ independently and randomly according to the distribution of $F(\cdot)$ from each of the n^d cells. We next independently select our sample of n cells as described above, letting

$$W_i = \begin{cases} 1, & \text{if cell } i \text{ is in our sample} \\ 0, & \text{otherwise} \end{cases}$$

and

$$G_i = g(y(\boldsymbol{X}_i)).$$

Then

$$\begin{aligned} \mathrm{Var}\{T_L\} &= \mathrm{Var}\left\{ \frac{1}{n} \sum_{j=1}^{n} G_j \right\} \\ &= \frac{1}{n^2} \left[\sum_{i=1}^{n^d} \mathrm{Var}\{W_i G_i\} \right. \\ &\qquad \left. + \sum_{i=1}^{n^d} \sum_{j=1, j \ne i}^{n^d} \mathrm{Cov}(W_i G_i, W_j G_j) \right]. \end{aligned}$$

To compute the variances and covariance on the right-hand side of this expression, we need to know some additional properties of the W_i. Using the fundamental rule that the probability of an event is the proportion of samples in which the event occurs, we find the following. First, $P(W_i = 1) = n/n^d = 1/n^{d-1}$ so W_i is Bernoulli with probability of success $1/n^{d-1}$. Second, if W_i and W_j correspond to cells having at least one common cell coordinate, then these two cells cannot both be selected, hence $\mathrm{E}\{(W_i W_j)\} = 0$. Third, if W_i and W_j correspond to cells having no cell coordinates in common, then

$$\mathrm{E}\{W_i W_j\} = P\{W_i = 1, W_j = 1\} = \frac{1}{n^{d-1}(n-1)^{d-1}}.$$

to answer analytically, and extensive empirical studies would be useful for better understanding what sorts of designs perform well and for which models.

6

Some Criterion-based Experimental Designs

6.1 Introduction

In Chapter 5, we considered designs that attempt to spread observations evenly throughout the experimental region. We called such designs space-filling designs. One rationale for using a space-filling design is the following. If we believe interesting features of the true model are just as likely to be in one part of the experimental region as another, we should take observations in all portions of the experimental region. A space-filling design attempts to do this. One difficulty is deciding exactly what it means for a set of observations to be evenly spread throughout the experimental region. There are many ways in which a design might be considered space-filling, and we discussed several in Chapter 5. Which design is best is not clear.

A more classical method of choosing a design is based on some statistical criterion, such as minimizing variances of estimators of model parameters. Suppose we intend to fit a second-order response surface model to the data resulting from a computer experiment. If we believe such a model adequately approximates the output of the computer code, we might select our design according to one of the many criteria proposed for second-order response surfaces. One such criterion is *D-optimality*; a D-optimal design minimizes the determinant of the covariance matrix of the least-square estimators of the regression parameters. This is equivalent to minimizing the volume of the confidence ellipsoid for the regression parameters. Other examples of criteria are: minimizing the integrated mean squared prediction error, and minimizing the squared bias (with respect to some true model).

In each case, we choose a design according to some statistical criterion. This is in contrast to Chapter 5 where we used geometric criteria that measured, in some sense, the spread of a collection of points.

In this chapter we consider some statistical criteria that have been used to construct experimental designs for computer experiments. As we will see, these are more difficult to implement than in the linear model because these criteria are functions of the unknown parameters for the Gaussian process models we have studied. Analytic results are difficult to obtain and have been found in only a few special cases. The details of such results are very technical and beyond the scope of this book. Here we describe the criteria and indicate how one might find good designs according to these criteria. This is an area of research in which much remains to be done.

We begin by discussing a criterion based on the notion of entropy and criteria based on the mean squared prediction error. Then we consider some sequential design strategies.

6.2 Designs Based on Entropy and Mean Squared Prediction Error Criterion

6.2.1 Maximum Entropy Designs

We begin by introducing the notion of entropy. Let X be a random variable taking a finite number of values which we take to be 1, 2, ..., n for simplicity. Let p_i be the probability that $X = i$. We define the *entropy* of X to be

$$H(X) = -\sum_{i=1}^{n} p_i \; \ell n \left(p_i\right), \tag{6.2.1}$$

where $\ell n(\cdot)$ denotes the natural logarithm and we define $p \times \ell n\,(p) = 0$ when $p = 0$. This definition can be extended to a continuous random variable having probability density function $f(\cdot)$ by defining the entropy of X to be

$$H(X) = -\int_{\mathcal{X}} f(x) \; \ell n \left(f(x)\right) dx \,.$$

What does entropy represent and why is this particular function used to measure entropy? Entropy is intended to be a measure of the unpredictability of a random variable. Intuitively, if all outcomes of a random variable X are equally likely, i.e., $p_1 = p_2 = \cdots = p_n = 1/n$, then X is maximally unpredictable and a reasonable definition of entropy should assign maximum value to this distribution. If X takes on a single value with probability 1, then X is completely predictable and should have minimum entropy. Thus, another way to think of entropy is as a measure of the uniformity and dispersion of the distribution of a random variable.

Notice that if X is a constant with probability one, then $H(X)$ defined by (6.2.1) is 0. This is the smallest possible value of $H(X)$ because $H(X) \geq 0$ for every X since all the p_i are between 0 and 1. Furthermore, it is not difficult to show for discrete X with n support points that the entropy is a maximum when $p_1 = p_2 = \cdots = p_n = 1/n$; in this case $H(X) = \ell n(n)$. Thus, $H(X)$ behaves as intuition suggests a measure of entropy should.

Lindley (1956) tied entropy to the amount of information contained in an experiment. His argument is based on one originally proposed by Shannon (1948). The basic idea is the following. The purpose of an experiment is to increase our information about a parameter $\boldsymbol{\theta}$ in a statistical model that we believe describes a measured response. For example, in regression, $\boldsymbol{\theta}$ might be the vector of regression parameters. In the Gaussian process models we consider in this book, $\boldsymbol{\theta}$ might represent the vector of correlation parameters. Knowledge about $\boldsymbol{\theta}$ is specified by a probability density or mass function $[\boldsymbol{\theta}]$ for $\boldsymbol{\theta}$. Here $[\boldsymbol{\theta}]$ could either be a prior distribution or a posterior distribution. How might we measure the information I about $\boldsymbol{\theta}$ that is contained in the distribution $[\boldsymbol{\theta}]$? Lindley describes I as the amount of information that must be provided to "know" $\boldsymbol{\theta}$. We take I to be a real-valued function of $[\boldsymbol{\theta}]$ and consider what properties I ought to have. For simplicity, we assume the set Θ of possible values of $\boldsymbol{\theta}$ is finite. One property is a consequence of considering a two-stage procedure for describing our knowledge about $\boldsymbol{\theta}$ that is equivalent to $[\boldsymbol{\theta}]$. Let Θ_1 be a nonempty proper subset of Θ with

$$0 < P \equiv \sum_{\boldsymbol{\theta} \in \Theta_1} [\boldsymbol{\theta}] < 1$$

and suppose in the first stage, we only consider whether $\boldsymbol{\theta} \in \Theta_1$ or its complement. The distribution that summarizes our knowledge of $\boldsymbol{\theta}$ at this stage is $(P, 1-P)$. This provides, say, amount I_1 of information. In the second stage, given $\boldsymbol{\theta} \in \Theta_1$, the distribution that summarizes the rest of our knowledge about $\boldsymbol{\theta}$ is $[\boldsymbol{\theta}]/P$. Given $\boldsymbol{\theta}$ is in the complement of Θ_1, the distribution that summarizes the rest of our knowledge about $\boldsymbol{\theta}$ is $[\boldsymbol{\theta}]/(1-P)$. The information provided in the second stage is, say, I_2 or I_3 according to whether $\boldsymbol{\theta} \in \Theta_1$ or $\boldsymbol{\theta}$ is in the complement of Θ_1. Then (as Shannon (1948) argues) one should require that the information provided in the first stage and the average (expected) amount of information provided in the second stage add up to the total information I in $[\boldsymbol{\theta}]$, namely

$$I = I_1 + PI_2 + (1-P)I_3.$$

This additivity requirement is the fundamental postulate for a measure of information I. Shannon (1948) showed that

$$I = \sum_{\Theta} [\boldsymbol{\theta}] \; \ell n\left([\boldsymbol{\theta}]\right)$$

is the only function having this property, apart from an arbitrary multiplying constant and a mild continuity property. As we do for entropy, we can extend this to continuous distributions by writing

$$I = \int_{\Theta} [\boldsymbol{\theta}] \; \ell n \left([\boldsymbol{\theta}] \right) d\boldsymbol{\theta} \, .$$

Information is typically viewed as the negative of a measure of entropy, namely $I = -H$. The above argument suggests that not only is $H(X)$ a reasonable measure of entropy, but it is the *only* measure of entropy having certain desirable properties.

Shewry and Wynn (1987) used these ideas to develop the notion of maximum entropy sampling when the design space is discrete. As above, let $[\boldsymbol{\theta}]$ be our prior distribution for $\boldsymbol{\theta}$, the parameter in our statistical model that we are interested in estimating. We observe a response at n input sites $\mathcal{D} = \{\boldsymbol{x}_1, \boldsymbol{x}_2, \ldots, \boldsymbol{x}_n\}$ and call $\mathcal{D}$ the experimental design. Let $[\boldsymbol{\theta} \mid \mathcal{D}]$ denote the posterior distribution of $\boldsymbol{\theta}$ given the data obtained using design $\mathcal{D}$. The amount of information in $[\boldsymbol{\theta}]$ about $\boldsymbol{\theta}$ before the experiment is

$$I = \int [\boldsymbol{\theta}] \; \ell n \left([\boldsymbol{\theta}] \right) d\boldsymbol{\theta}$$

and the amount of information about $\boldsymbol{\theta}$ after experiment using design $\mathcal{D}$ is

$$I_{\mathcal{D}} = \int [\boldsymbol{\theta} \mid \mathcal{D}] \; \ell n \left([\boldsymbol{\theta} \mid \mathcal{D}] \right) d\boldsymbol{\theta}.$$

Thus, the change in information is $I - I_{\mathcal{D}}$. To evaluate a design, we consider the expected change. Using the fact that entropy is the negative of information, Shewry and Wynn (1987) showed that the expected change in information is maximized by the design $\mathcal{D}$ that maximizes the entropy of the observed responses at the points in the design. Such a design is called a *maximum entropy design.*

In the case of the Gaussian process models we are considering, recall that the training data has the conditional distribution

$$\boldsymbol{Y}^n \mid \boldsymbol{\beta} \sim N_n(\boldsymbol{F}\boldsymbol{\beta}, \sigma_Z^2 \boldsymbol{R}).$$

One can show that a maximum entropy design maximizes the determinant of the variance of the observed responses at the points in the design. If we assume the Gaussian prior is

$$\boldsymbol{\beta} \sim N_p \left(\boldsymbol{b}_0, \tau^2 \boldsymbol{V}_0 \right),$$

The IMSPE and MMSPE design criteria are generalizations of the classical A-optimality and G-optimality criteria. To see this, suppose that $Z(\boldsymbol{x})$ in (2.3.6) is a *white noise process.* This is the traditional regression setting with random error used in response surface modeling for physical experiments.

From (6.2.2) the IMSPE criterion function (6.2.3) simplifies to

$$\begin{aligned} J(\mathcal{D}, \widehat{Y}) - 1 &= \int_{\mathcal{X}} \boldsymbol{f}^\top(\boldsymbol{x})(\boldsymbol{F}^\top \boldsymbol{F})^{-1} \boldsymbol{f}(\boldsymbol{x})\, w(\boldsymbol{x})\, d\boldsymbol{x} \\ &= \operatorname{tr}\left(\boldsymbol{W}(\boldsymbol{F}^\top \boldsymbol{F})^{-1}\right), \end{aligned}$$

where

$$\boldsymbol{W} = \int_{\mathcal{X}} \boldsymbol{f}(\boldsymbol{x}) \boldsymbol{f}^\top(\boldsymbol{x})\, w(\boldsymbol{x})\, d\boldsymbol{x}$$

is a positive definite $p \times p$ weight matrix. A design minimizing

$$\operatorname{tr}\left(\boldsymbol{W}(\boldsymbol{F}^\top \boldsymbol{F})^{-1}\right)$$

is called *L-optimal* (see Pukelsheim (1993)). Thus, the IMSPE-optimal designs are the L-optimal designs when $Z(\boldsymbol{x})$ is a white noise process.

L-optimal designs are related to A-optimal designs as follows. Let $\boldsymbol{H}$ be a $p \times p$ square root of $\boldsymbol{W}$; that is, $\boldsymbol{W} = \boldsymbol{H}\boldsymbol{H}^\top$. Then,

$$\operatorname{tr}\left(\boldsymbol{W}(\boldsymbol{F}^\top \boldsymbol{F})^{-1}\right) = \operatorname{tr}\left(\boldsymbol{H}^\top (\boldsymbol{F}^\top \boldsymbol{F})^{-1} \boldsymbol{H}\right)$$

so a design is L-optimal if and only if it is *A-optimal* for the parameter subsystem $\boldsymbol{H}^\top \boldsymbol{\beta}$.

The MMSPE criterion function (6.2.6) simplifies to

$$Q(\mathcal{D}, \widehat{Y}) - 1 = \max_{\boldsymbol{x} \in \mathcal{X}} \boldsymbol{f}^\top(\boldsymbol{x})(\boldsymbol{F}^\top \boldsymbol{F})^{-1} \boldsymbol{f}(\boldsymbol{x}).$$

A design minimizing

$$\max_{\boldsymbol{x} \in \mathcal{X}} \boldsymbol{f}^\top(\boldsymbol{x})(\boldsymbol{F}^\top \boldsymbol{F})^{-1} \boldsymbol{f}(\boldsymbol{x})$$

is called *G-optimal.* Thus, the MMSPE-optimal designs are the G-optimal designs when $Z(\boldsymbol{x})$ is a white noise process.

Our experience is that, because of a lack of tabulated designs and a lack of easily accessible software for constructing designs, maximum entropy, IMSPE-optimal, and MMSPE-optimal designs are not often used by practitioners. LHDs, which are relatively easy to construct for any sample size, continue to be a more popular choice.

6.3 Designs Based on Optimization Criteria

6.3.1 Introduction

This section describes several sequential experimental design strategies that have been proposed for the important problem of finding inputs $\boldsymbol{x}$ that optimize the output of a computer code. Given input vector $\boldsymbol{x} \in \mathcal{X}$, assume that the code calculates one or more responses $y_1(\boldsymbol{x}), \ldots, y_m(\boldsymbol{x})$ $(m \geq 1)$. In some applications, the $y_i(\cdot)$ are the objects of direct interest while in other applications, either integrals or linear combinations of the $y_i(\cdot)$ are of interest. As illustrated in Section 1.2, an example of the latter occurs when $\boldsymbol{x}$ consists of both control and environmental components, i.e., $\boldsymbol{x} = (\boldsymbol{x}_c, \boldsymbol{x}_e)$. In this case, suppose that the environmental variables have a *known* probability distribution, which is specified either by the probability mass function $w_j = P\{\, \boldsymbol{X}_e = \boldsymbol{x}_{e,j} \,\}$ on n_e support points $\{\boldsymbol{x}_{e,j}\}$ or by the probability density function $w(\cdot)$. Then the quantities of interest are

$$\mu_i(\boldsymbol{x}_c) \equiv \sum_{j=1}^{n_e} w_j \, y_i(\boldsymbol{x}_c, \boldsymbol{x}_{e,j}) \quad \left(\text{or } \equiv \int y_i(\boldsymbol{x}_c, \boldsymbol{x}_e) \, w(\boldsymbol{x}_e) \, d\boldsymbol{x}_e \right), \tag{6.3.1}$$

which is the mean of $y_i(\cdot)$ with respect to the distribution of the $\boldsymbol{X}_e$ variables for $i = 1, \ldots, m$.

Subsections 6.3.2–6.3.5 consider problems of optimizing a *single* output $y_1(\cdot)$ $(m = 1)$. We consider two specific problems of this sort. The first is that of minimizing $y_1(\boldsymbol{x})$ as a function of *all* the input variables $\boldsymbol{x}$. When $\boldsymbol{x}$ consists of control and environmental components, a second problem is that of finding control variable combinations $\boldsymbol{x}_{c,\min}$ that minimize $\mu_1(\cdot)$. When there are multiple outputs $(m \geq 2)$, Subsection 6.3.6 describes a method of locating a minimizer $\boldsymbol{x}_{\min}$ of $y_1(\cdot)$ that satisfies feasibility constraints on $y_2(\cdot), \ldots, y_m(\cdot)$, and also a method for finding control variable combinations $\boldsymbol{x}_{c,\min}$ that minimize $\mu_1(\cdot)$, subject to feasibility constraints on $\mu_2(\cdot), \ldots, \mu_m(\cdot)$.

The optimization algorithms presented in this section utilize multiple experimental design stages. The idea of these methods is to use first-stage data to obtain initial information about the *entire* response surface, while each additional stage takes account of all previous information to obtain an experimental design consistent with the optimization objective. A quantitative criterion is implemented by each algorithm for the purpose of deriving the experimental design in each stage.

6.3.2 Bernardo Multi-Stage Approximation

Bernardo et al. (1992) proposed a sequential strategy for optimizing integrated circuit designs. Conceptually, their algorithm should be thought of as minimizing a single function, $y_1(\cdot)$, over inputs $\boldsymbol{x} \in \mathcal{X}$; thus this is an $m = 1$ problem. Their method sequentially refines the region of the input space where an optimum appeared to be located. In common with the example discussed in Subsection 1.2.2, computational considerations limited the number of model runs they could make and necessitated the use of a surrogate predictor for model output. In overview, the algorithm is implemented as follows.

1. Postulate an approximating model for the data generating process.

Bernardo et al. (1992) adopted the stochastic process model of Section 2.3, with power exponential correlation function, which we have seen in Section 3.3 is substantially more flexible than standard polynomial models and implicitly accounts for nonlinearities in the inputs and complex interactions.

2. Design an initial experiment and collect the required data. Estimate model parameters and calculate the response predictor.

Bernardo et al. (1992) recommended initial designs containing *at least three observations* per *estimated model parameter* (in contrast, Jones et al. (1998) recommended larger initial designs having ten observations per *input variable*). Latin hypercube designs were run at the initial and each subsequent stage of this algorithm. As we have seen in Subsection 5.2.2, designs based on Latin hypercube samples have attractive marginal projection properties, while Section 5.5 shows that maximin distance LHDs provide a more uniform distribution of points for higher dimensional projections. Bernardo et al. (1992) used an empirical BLUP based on their stochastic process model as a computationally inexpensive surrogate for the true model output.

3. Check prediction accuracy and visualize the fitted models. If the prediction accuracy is sufficient, predict the global optimizer. Otherwise, go to Step 4.

Prediction accuracy is judged by computing the empirical root mean square prediction error (see (4.2.13)) for a set of randomly selected points in the analysis region and also by examining the range of predicted values. Thus this assessment of prediction accuracy depends on the particular application. In their circuit design example, Bernardo et al. (1992) compared the ERMSPE to the maximum allowable limit of variability in the reference current. In general, prediction accuracy is measured by the values of the ERMSPE relative to typical values of the response as dictated by the particular physical application under study.

4. Choose a subregion for the next experiment, and go to Step 1.

If additional stages are required, they are run on a *subregion* of the previous stage that contains the current *predicted* optimum. Sensitivity analysis methodology described in Section 7.1 provides a mechanism for quantifying the importance of input variables. In particular, main effect and interaction plots provide guidance for supplying reasonable input variable ranges to the next stage. Inactive inputs can be set equal to nominal values, which reduces the dimension of the input variable space in subsequent stages. For model fitting, Bernardo et al. (1992) discarded data from previous stages for inputs falling outside the subregion of the current stage. Once the response predictor attains the accuracy requirement, a confirmatory run is made at the location of the predicted optimum. If the actual results at the predicted optimum violate problem specifications, adjustments are made to the statistical model and the algorithm continues until an acceptable solution is obtained. For example, in their integrated circuit study, Bernardo et al. (1992) added linear regression terms to the statistical model in response to a failed confirmatory run. This led to increased prediction accuracy and a successful confirmatory run terminating the algorithm.

6.3.3 The Surrogate Management Framework Algorithm

Booker et al. (1999) implemented a "direct search" method called the *Surrogate Management Framework* (SMF) for minimizing the output $y_1(\cdot)$ of an expensive computer model, where the inputs are assumed to be located in a rectangular region $\mathcal{X}$. Their SMF algorithm is based on an optimization algorithm called the *Generalized Pattern Search* (GPS) algorithm, several of whose features we describe next.

Several terms must be introduced to sketch the GPS algorithm. A collection of vectors forms a *positive basis* for $\mathbb{R}^d$ if nonnegative linear combinations of these vectors span $\mathbb{R}^d$, and no proper subset also forms a spanning set. Unlike the usual basis of a vector space, the number of vectors in a positive basis for $\mathbb{R}^d$ is not unique; there are $d+1$ vectors in a minimal-sized positive basis, and $2d$ vectors in a maximal-sized positive basis.

After each new input has been determined, the GPS algorithm must have a well-defined set of input variable locations from which the next iterate is to be selected—this is called a *mesh.* The mesh must be constructed in such a way that each element $\boldsymbol{x}_0$ of the mesh has a set of neighbors for which the collection of differences between these neighbors and $\boldsymbol{x}_0$ contains a positive basis for $\mathbb{R}^d$. This condition ensures that at least one of these difference vectors in the positive basis is a direction of descent at $\boldsymbol{x}_0$, provided the gradient of $y_1(\cdot)$ at $\boldsymbol{x}_0$ is not equal to zero. This fact leads to desirable convergence properties of the GPS.

For $n = 0, 1, \ldots$, let M_n denote the mesh on $\mathcal{X}$ corresponding to the nth iteration of the GPS, and let $\boldsymbol{x}_n \in M_n$ denote the input at the nth iterate. If $\boldsymbol{x}_n$ is in the interior of M_n, construct a subset X_n containing $\boldsymbol{x}_n$ and a set of $2d$ inputs in M_n adjacent to $\boldsymbol{x}_n$ for which the differences between these inputs and $\boldsymbol{x}_n$ form a maximal positive basis for $\mathbb{R}^d$. If $\boldsymbol{x}_n$ is on the boundary of M_n, then one constructs a positive basis with fewer elements. By construction, M_n contains the required points to form such a

estimated by maximum likelihood or restricted maximum likelihood, say, as described in Subsection 3.3.2.

These algorithms are based on a notion of "improvement" that is defined as follows. Let

$$y^n_{\text{min}} = \min_{i=1,\ldots,n} y_1(\boldsymbol{x}_i)$$

denote the minimum output that has been evaluated after n runs of the computer code. Consider a *potential* site $\boldsymbol{x}$ at which to evaluate the code. Compared with the current smallest *known* minimum value of $y_1(\cdot)$, we define the amount of improvement in $y_1(\cdot)$ at $\boldsymbol{x}$ to be *zero* if $y_1(\boldsymbol{x}) \geq y^n_{\text{min}}$, i.e., $y_1(\boldsymbol{x})$ provides no improvement over y^n_{min}. Similarly if $y_1(\boldsymbol{x}) < y^n_{\text{min}}$, the amount of improvement at $\boldsymbol{x}$ is defined to be the difference $y^n_{\text{min}} - y_1(\boldsymbol{x})$. Hence, in principle, we define the improvement possible at $\boldsymbol{x}$ to be

$$\text{Improvement at } \boldsymbol{x} = \begin{cases} y^n_{\text{min}} - y_1(\boldsymbol{x}), & y^n_{\text{min}} - y_1(\boldsymbol{x}) > 0 \\ 0, & y^n_{\text{min}} - y_1(\boldsymbol{x}) \leq 0 \end{cases}.$$

We say "in principle" because $y_1(\boldsymbol{x})$ is *unknown*, although y^n_{min} is known from the training data. However, we have an idea of the location of $y_1(\boldsymbol{x})$ from its prior $Y_1(\boldsymbol{x})$ (and a better one from the updated prior—the posterior of $Y_1(\boldsymbol{x})$ given $\boldsymbol{Y}^n_1$ in (6.3.3)). Hence a probabilistically-based *improvement function* can be defined by

$$I_n(\boldsymbol{x}) = \begin{cases} y^n_{\text{min}} - Y_1(\boldsymbol{x}), & y^n_{\text{min}} - Y_1(\boldsymbol{x}) > 0 \\ 0, & y^n_{\text{min}} - Y_1(\boldsymbol{x}) \leq 0 \end{cases} \tag{6.3.5}$$

for $\boldsymbol{x} \in \mathcal{X}$. The random $I_n(\boldsymbol{x})$ depends solely on the random quantity $Y_1(\boldsymbol{x})$. We summarize the amount of improvement possible at each potential input site $\boldsymbol{x}$ by its *expected improvement* with respect to the posterior distribution of $[Y_1(\boldsymbol{x})|\boldsymbol{Y}^n_1]$.

Example 6.2 Suppose that the input space $\mathcal{X}$ is one-dimensional and that Figure 6.1 shows the posterior densities of $[\, y^n_{\text{min}} - Y_1(x) \,|\, \boldsymbol{Y}^n_1 = \boldsymbol{y}^n_1 \,]$ for $x \in \{x_1, x_2, x_3\}$. Examining the conditional density of $y^n_{\text{min}} - Y_1(x_1)$ shows that x_1 is a good candidate for exploration because this density is concentrated on positive values. This fact suggests that $y_1(x_1)$ is likely to be lower than y^n_{min}. On the other hand, x_2 is not a promising site for exploration because the posterior density of $y^n_{\text{min}} - Y_1(x_2)$ is concentrated on negative values. Finally, note that the posterior density of $y^n_{\text{min}} - Y_1(x_3)$ has relatively heavy tails and much of the support of the density is again over the positive numbers. Thus, x_3 is also a reasonable candidate for exploration. The heavy tails of this last density are a result of the fact that the MSPE at site x_3, $s^2_1(x_3)$ from (6.3.4), is large. Taking another observation at x_3 would decrease the conditional MSPE $s^2_1(x_3)$ given all $n+1$ outputs. In general, we will see below that input sites for which $s^2_1(\boldsymbol{x})$ is large can yield large

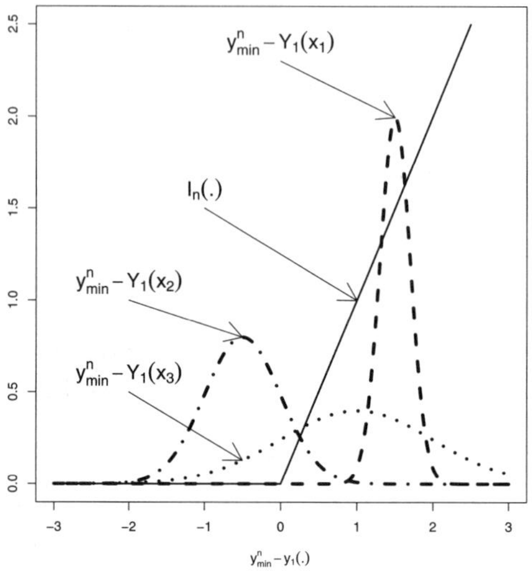

FIGURE 6.1. The conditional (posterior) density of $y^n_{\rm min} - Y_1(x)$ given $\boldsymbol{Y}^n_1$ for $x \in \{x_1, x_2, x_3\}$. The improvement function $I_n(\cdot)$ is plotted as a solid line.

values of the *conditional expected improvement.* In particular, there is the potential for substantial improvement by investigating x_3. ■

It is straightforward to show that the *conditional expected improvement* satisfies $\mathrm{E}\{\, I_n(\boldsymbol{x}) \,|\, \boldsymbol{Y}^n_1 \,\} = 0$ for $\boldsymbol{x}$ in the input training data $\mathcal{D}_n$. This result coincides with our intuition that there is no benefit in recomputing the output at sites $\boldsymbol{x}$ where we know $y_1(\boldsymbol{x})$. If $\boldsymbol{x} \notin \mathcal{D}_n$, some algebra shows that

$$\mathrm{E}\{\, I_n(\boldsymbol{x}) \,|\, \boldsymbol{Y}^n_1 \,\} =$$
$$s_1(\boldsymbol{x}) \left\{ \frac{y^n_{\rm min} - \widehat{Y}_1(\boldsymbol{x})}{s_1(\boldsymbol{x})} \, \Phi\left(\frac{y^n_{\rm min} - \widehat{Y}_1(\boldsymbol{x})}{s_1(\boldsymbol{x})} \right) + \phi\left(\frac{y^n_{\rm min} - \widehat{Y}_1(\boldsymbol{x})}{s_1(\boldsymbol{x})} \right) \right\}, \tag{6.3.6}$$

where $\Phi(\cdot)$ and $\phi(\cdot)$ are the $N(0,1)$ distribution and density function, respectively. By examining the terms in (6.3.6), we see that the posterior expected improvement is "large" for those $\boldsymbol{x}$ having either

- a predicted value at $\boldsymbol{x}$ that is much smaller than the best minimum computed so far, i.e., $\widehat{Y}_1(\boldsymbol{x}) \ll y^n_{\rm min}$, *or*
- having much uncertainty about the value of $y_1(\boldsymbol{x})$, i.e., when $s_1(\boldsymbol{x})$ is large.

These observations quantify the discussion in Example 6.2 about the operation of the algorithm. Candidate inputs are judged attractive if *either* there is high probability that their predicted output is below the current observed minimum *and/or* there is a large uncertainty in the predicted output.

Starting with a space-filling design, the expected improvement algorithm updates the current input set $\mathcal{D}_n$ as follows.

Given the specified absolute tolerance ϵ_a, *if*

$$\max_{\boldsymbol{x} \in \mathcal{X}} E\{ I_n(\boldsymbol{x}) \mid \boldsymbol{Y}_1^n \} < \epsilon_a, \tag{6.3.7}$$

then stop and predict $\boldsymbol{x}_{\min}$ *by an input site* $\widehat{\boldsymbol{x}}_{\min} \in \{\boldsymbol{x}_1, \ldots, \boldsymbol{x}_n\}$ *satisfying*

$$y_1(\widehat{\boldsymbol{x}}_{\min}) = \min_{i=1,\ldots,n} y_1(\boldsymbol{x}_i)\,.$$

Otherwise, select the point $\boldsymbol{x}_{n+1} \in \mathcal{X}$ *to maximize* $E\{ I_n(\boldsymbol{x}) \mid \boldsymbol{Y}_1^n \}$. *Set* $\mathcal{D}_{n+1} = \mathcal{D}_n \cup \{\boldsymbol{x}_{n+1}\}$, $\boldsymbol{Y}_1^{n+1} = ((\boldsymbol{Y}_1^n)^\top, y_1(\boldsymbol{x}_{n+1}))^\top$, *and increment* n. *Continue with the next update.*

Among the obvious modifications of this algorithm are the following. At stopping, one can predict $\boldsymbol{x}_{\min}$ to be the minimizer of the predictor $\widehat{Y}_1(\cdot)$ (based on the data available). The absolute stopping criterion (6.3.7) can be replaced by the relative stopping criterion

$$\frac{\max_{\boldsymbol{x} \in \mathcal{X}} \mathrm{E}\{ I_n(\boldsymbol{x}) \mid \boldsymbol{Y}_1^n \}}{|y_{\min}^n|} < \epsilon_r,$$

where ϵ_r is a specified relative tolerance. In either case of the stopping criterion, the idea is that the algorithm should continue until the maximum possible improvement is small.

Jones et al. (1998) (and Schonlau et al. (1998)) adopted the power exponential correlation structure of (2.3.14) for the Gaussian process $Y_1(\cdot)$ with one range and one smoothness parameter per input. They estimate these parameters by the method of maximum likelihood. Upon completion of each update step, the correlation parameters of the stochastic model can optionally be updated. The updating procedure can be computationally expensive, particularly for large designs. Schonlau et al. (1998) provided a modification of the expected improvement algorithm that accommodates generating several input sites at each stage, with correlation parameter updates taking place at the completion of each stage. Specifically, given a current design of size n and q iterates to be added, Schonlau et al. (1998) recommended updating the $s_1(\boldsymbol{x})$ coefficient in (6.3.6) after each iterate, but not updating the $s_1(\boldsymbol{x})$ term in the expressions $(y_{\min}^n - \widehat{Y}_1(\boldsymbol{x}))/s_1(\boldsymbol{x})$. This heuristic forces all previously sampled inputs to be avoided, including the previous iterates of the current stage, as $s_1(\cdot)$ is 0 at these inputs. The empirical BLUP and MSPE are updated subsequent to the correlation parameters at the completion of each stage. If $\mathcal{X}$ is finite, the expected improvement algorithm will converge to the global minimum under the assumption that ϵ_a (or ϵ_r) $= 0$. Schonlau (1997) demonstrated the effectiveness of this algorithm for a suite of test problems where ϵ_a (or ϵ_r) > 0.

Schonlau et al. (1998) considered a generalization of the expected improvement criterion in which the improvement (6.3.5) is replaced by

$$I_n^g(\boldsymbol{x}) = \begin{cases} (y_{\min}^n - Y_1(\boldsymbol{x}))^g, & \text{if } Y_1(\boldsymbol{x}) < y_{\min}^n \\ 0, & \text{otherwise} \end{cases}$$

for some $g \in \{0, 1, 2, \ldots\}$. Larger values of g are associated with a more *global* search. This can be seen by examining Figure 6.1. Provided $I_n(\boldsymbol{x}) \geq 1$, $I_n^{g_1}(\boldsymbol{x}) \geq I_n^{g_2}(\boldsymbol{x})$ for each input $\boldsymbol{x}$ when $g_1 \geq g_2$. Therefore, greater weight will be placed on the tails of the conditional distribution of $Y_1(\cdot)$ given $\boldsymbol{Y}_1^n$ for larger values of g, so that the global potential for large improvement is given increased quantitative importance. The quantity $\mathrm{E}\{\, I_n(\cdot) \,|\, \boldsymbol{Y}_1^n \,\}$ in the stopping rule for the expected improvement algorithm is replaced by $\mathrm{E}\{\, I_n^g(\cdot) \,|\, \boldsymbol{Y}_1^n \,\}^{1/g}$ so that the tolerance limits ϵ_a and ϵ_r have approximately the same interpretation for any g. Schonlau et al. (1998) provided recursive relations for computing $\mathrm{E}\{\, I_n^g(\boldsymbol{x}) \,|\, \boldsymbol{Y}_1^n \,\}$.

Unconstrained Optimization II

Williams et al. (2000) extended the expected improvement algorithm to settings involving both control and environmental inputs. For discrete environmental variable distributions as used in the left-hand side of (6.3.1), the goal of their revised algorithm was to find a control variable input that minimized the mean function $\mu_1(\cdot)$ for applications with "expensive" $y_1(\cdot)$ outputs. Specifically, denote the jth support point and weight of the environmental probability distribution by $\boldsymbol{x}_{e,j}$ and $\{w_j\}$, $j = 1, \ldots, n_e$. The mean $\mu_1(\boldsymbol{x}_c)$ is the output $y_1((\boldsymbol{x}_c, \boldsymbol{x}_{e,j}))$ averaged over these n_e values. The function $\mu_1(\boldsymbol{x}_c)$ inherits the prior process defined by

$$M_1(\boldsymbol{x}_c) \equiv \sum_{j=1}^{n_e} w_j \, Y_1((\boldsymbol{x}_c, \boldsymbol{x}_{e,j})),$$

where $Y_1((\boldsymbol{x}_c, \boldsymbol{x}_e))$ has the Gaussian prior used by Schonlau et al. (1998) and Jones et al. (1998).

Let $\mathcal{D}_n = \{(\boldsymbol{x}_{c,i}, \boldsymbol{x}_{e,i}) \,,\ 1 \leq i \leq n\}$ denote a generic n-point experimental design. Then $\{\boldsymbol{x}_{c,i}\}$ denotes the control variable portion of this design. Williams et al. (2000) extended the improvement function (6.3.5) to this setting by defining their improvement function to be

$$I_n(\boldsymbol{x}_c) = \begin{cases} M_{\min}^n - M_1(\boldsymbol{x}_c), & M_{\min}^n - M_1(\boldsymbol{x}_c) > 0 \\ 0, & M_{\min}^n - M_1(\boldsymbol{x}_c) \leq 0 \end{cases},$$

choosing the environmental variable site for the subsequent iteration of the algorithm. Let $\boldsymbol{Y}_1^{n_1}$ and $\boldsymbol{Y}_2^{n_2}$ denote the vectors of outcomes calculated on the current experimental designs of n_1 and n_2 sites for the first and second outcomes, respectively, and let $n = n_1 + n_2$ denote the current total number of experimental design sites. Set $\boldsymbol{Y}^n = (\boldsymbol{Y}_1^{n_1}, \boldsymbol{Y}_2^{n_2})$, $\boldsymbol{Y}_e^j = (Y_j(\boldsymbol{x}_{c,n+1}, \boldsymbol{x}_e), \boldsymbol{Y}^n)$ and let $\widehat{M_i^j}(\boldsymbol{x}_c)$ denote the conditional mean of $M_i(\boldsymbol{x}_c)$ given $\boldsymbol{Y}_e^j$. The control variable site that maximizes conditional expected improvement given the current n-point experimental design is designated $\boldsymbol{x}_{c,n+1}$. For $j = 1, 2$, define the "mean squared prediction error" function by

$$\begin{aligned}\mathrm{MSPE}_j(\boldsymbol{x}_e) = \mathrm{E}\{\, [\widehat{M_1^j}(\boldsymbol{x}_{c,n+1}) - M_1(\boldsymbol{x}_{c,n+1})]^2 \,|\, \boldsymbol{Y}^n \,\} + \\ \mathrm{P}\{\, M_2(\boldsymbol{x}_{c,n+1}) > u_2 \,|\, \boldsymbol{Y}^n \,\}\, \mathrm{E}\{\, [\widehat{M_2^j}(\boldsymbol{x}_{c,n+1}) - M_2(\boldsymbol{x}_{c,n+1})]^2 \,|\, \boldsymbol{Y}^n \,\} .\end{aligned}$$

For each outcome j, calculate

$$\boldsymbol{x}_{e,j}^* = \arg \min_{\boldsymbol{x}_e \in \mathcal{X}_e} \mathrm{MSPE}_j(\boldsymbol{x}_e) .$$

If $\mathrm{MSPE}_1(\boldsymbol{x}_{e,1}^*) \leq \mathrm{MSPE}_2(\boldsymbol{x}_{e,2}^*)$, augment the experimental design corresponding to the first outcome with $(\boldsymbol{x}_{c,n+1}, \boldsymbol{x}_{e,1}^*)$ and compute $y_1(\cdot)$ at this location; otherwise, augment the experimental design corresponding to the second outcome with $(\boldsymbol{x}_{c,n+1}, \boldsymbol{x}_{e,2}^*)$ and compute $y_2(\cdot)$ at this location. This criterion selects the next outcome to be generated based on minimizing a weighted sum of prediction errors in the objective and constraint functions. The error in predicting the constraint function is weighted by the probability that the constraint is exceeded given the current data; if this probability is low, then constraint satisfaction is currently determined with certainty and minimizing prediction error in the constrained outcome is unnecessary. On the other hand, if constraint satisfaction is less certain, the need to reduce prediction error in the objective function is balanced against the desire to reduce prediction uncertainty in the constraint function.

7

Sensitivity Analysis, Validation, and Other Issues

7.1 Sensitivity Analysis

7.1.1 *Introduction*

In this section we discuss sensitivity analysis. In general, sensitivity analysis is the study of how variation in an observed response can be apportioned to different possible sources or factors. In computer experiments, the "observed response" is the output of the computer code and the "factors" are inputs to the code. In other words, sensitivity analysis tries to determine how variable the output is to changes in the inputs. There are many applications of sensitivity analysis and Saltelli, Chan and Scott (2000) discuss these in detail. One application of sensitivity analysis is the following. Suppose we collect data to determine whether a response depends on any of several factors. We wish to identify which factors are responsible for most of the variation in the response and which produce little variation in the response over some specified range of values of these factors. Sensitivity analysis provides methods that can be used to accomplish this.

In computer experiments, we know a priori that the inputs we provide to the code affect the response. These inputs are part of the mathematical description of the physical process that forms the basis for the code. However, over the experimental region we are interested in exploring, some of the inputs may have relatively *little* effect on the output, i.e., the output may not be sensitive to changes in these inputs. When this is the case, we can set these inputs to some nominal value (for example, their mean if the variables are noise variables) and investigate how the output depends on the remaining inputs. This reduces the dimensionality of the problem and allows us to fit a model that uses fewer inputs. As a consequence, we may require fewer observations in the experimental region to adequately fit a statistical model to the output and, if so, reduce the possibility of numerical problems when fitting a model to the data. Also recall that in (engineering) applications having both control and noise variables, the output is considered "robust" to those noise variables for which it is insensitive.

Sensitivity analysis is also useful for identifying interactions between variables. When interactions do not exist, the effect of any given input is the same regardless of the values of the other inputs. In this case, the relation between the output and inputs is said to be additive and is relatively simple to understand. When interactions exist, the effects of some inputs on the output will depend on the values of other inputs. There are circumstances in which the presence of interactions allows the experimenter to attain important practical goals; for example, it may be the case that for certain settings of some of the inputs, the output is insensitive to other variables. If these other variables are noise variables, we can use the first set of inputs to make the output "robust" to these noise variables.

In computer experiments, a sensitivity analysis can be applied either to the output of the code itself (the training data) or to a predictor that we fit to the data (an EBLUP, for example). In the former case, it is assumed that additonal observations are "easy" to generate. In the latter case, observations are expensive or time-consuming to compute and relatively few observations are available for the analysis. *But* in the latter case, our conclusions technically apply to the fitted model rather than to the code itself. Inferring that our conclusions also apply to the code is only valid to the extent that our predictor fits the output of the code well.

How can one carry out a sensitivity analysis? For a complete discussion, one should refer to the book length description in Saltelli et al. (2000). We content ourselves with brief descriptions of some of the many possible analyses.

7.1.2 Sensitivity Analyses Based on Scatterplots and Correlations

A simple approach is the following. Make a scatterplot of each input versus the output of the code and from these plots, determine those inputs that appear to produce large variation in the output and those which produce small variation. The correlations between each input and the output can also be computed, but correlation coefficients *only* indicate the extent to which there is *linear association* between the outputs and the input. Scatterplots are generally more informative than correlations because nonlinear relationships can be seen in plots, whereas correlations only indicate the presence of straight-line relationships.

As an example, Figure 1.6 on page 11 plots the failure depth of pockets punched into sheet metal (the output) versus clearance and versus fillet radius, two characteristics of the machine tool used to form the pockets. The scatterplot of failure depth versus clearance shows an increasing trend, suggesting that failure depth is sensitive to clearance. However, in the scatterplot of failure depth versus fillet radius, no trend appears to be present, suggesting that failure depth may not be sensitive to fillet radius.

One thing that simple scatterplots do not indicate is whether interaction effects exist. Three-dimensional plots (of the output versus pairs of inputs), two-dimensional interaction plots (also called "mean plots" or "profile plots") that use different plotting symbols to represent inputs that have only a few values, or a series of two-dimensional plots that condition on one or more inputs (called "trellis plots" in S-PLUS, Cleveland (1993)), can all be used to explore two-factor interaction effects (but not higher-order effects).

7.1.3 Sensitivity Analyses Based on Regression Modeling

Another approach is to fit regression models to the output and to assess the "importance" of input variables by their regression coefficients. Such methods are most effective when the design is orthogonal or nearly orthogonal.

One popular procedure that uses this strategy is the following. Standardize all the variables (both the output and the inputs) by subtracting from each its mean and dividing the result by the sample standard deviation of the variable. For example, suppose input x_i has values $x_{i_1}, x_{i_2}, \ldots, x_{i_n}$. Let $\overline{x}_i$ denote the mean of these values and s_{x_i} their standard deviation. Then, the standardized value of x_{i_j} is

$$\frac{x_{i_j} - \overline{x}_i}{s_{x_i}}.$$

Using standardized values places all variables on a common scale. Having standardized all variables, fit the first-order regression model to the standardized data. The regression coefficients are called the *standardized regression coefficients* (SRCs). Because all variables have been placed on a common scale, the relative magnitudes of the SRCs indicate the relative sensitivity of the output to each input. The output is most sensitive to those inputs whose SRCs are largest in absolute value. The validity of this interpretation depends on the overall fit of the regression model, either as indicated by measures such as the coefficient of determination R^2, or by predicted sums of squares. If the overall fit is poor, the regression model does not adequately describe the relation between the output and input, and the SRCs do not reflect the effect of the inputs on the output.

Partial correlation coefficients (PCCs) between the output and the inputs are also sometimes used to assess sensitivity. PCCs measure the strength of the linear relationship between the output and a given input, after adjusting for any linear effects of the other inputs. The relative sizes of PCCs are used to assess the sensitivity of the output to the inputs. Again, if the overall fit of the model is poor, PCCs do not give useful information about the sensitivity of the output to the inputs. Also, both SRCs and PCCs become difficult to interpret if a high degree of collinearity is present.

A variation on the above approach is to first rank transform the data (both the inputs and the output), and then fit a first-order regression model to the transformed data. The rank transformation is carried out as follows. For each variable, rank the observed values from low to high. If there are N values, the lowest is given rank 1, the next lowest rank 2, etc., and the highest rank N. Use the average rank for ties. One replaces the original data by these ranks and fits a first-order regression model to the ranked data. Standardized regression coefficients or partial correlations are used to assess the sensitivity of the output to the inputs. Once again, if the overall fit of the first-order regression model is poor, the standardized regression coefficients or partial correlations do not adequately describe the effect of the inputs on the output and this analysis does not provide good information about sensitivities.

In practice, it has been observed that the regression model for the ranked transformed data often has higher R^2 values than that for the regression model based on the standardized data. This may be because the rank transformation removes certain nonlinearities present in the original data. Thus, when monotone (but nonlinear) trends are present, there are some advantages to conducting a sensitivity analysis using the rank transformed data. However, when one uses the rank transformed data, one must keep in mind that the resulting measures of sensitivity give us information on the sensitivity of the ranked transformed output to the rank transformed inputs, rather than on the original variables.

Whether one uses the standardized data or the rank transformed data, because we are fitting first-order models, we get no information on interactions or on nonmonotone effects of variables. If one has reason to believe that interactions are present, or that the relation between the output and some of the inputs is nonlinear and nonmonotone, these regression methods will not give reliable information about sensitivities. One may wish to consider fitting higher-order models in such cases.

Stepwise regression is another method for constructing a regression model and one that can be used to assess sensitivity. For example, if a forward stepwise regression is used, the first variable entered would be considered the most influential input, the second variable entered would be considered the second most influential input, etc. As is usual in stepwise regression, one continues until the amount of variation explained by adding further variables is not considered meaningful according to some criterion selected by the user. Statistics such as the mean squared error, the F-statistic for testing whether the addition of another variable significantly improves the model, the coefficient of determination R^2, or the adjusted R^2 can be used to determine when to stop the stepwise regression. For more on stepwise regression, see any standard text on regression, for example Draper and Smith (1981).

Rather than fitting first-order models, one can fit a second-order response surface to the output. This allows one to explore second-order (quadratic) effects of inputs and two-factor interaction (cross-product) effects. For more on response surface methods, see Box and Draper (1987).

7.1.4 Sensitivity Analyses Based on ANOVA-Type Decompositions

Sobol´ (1990), Welch et al. (1992), and Sobol´ (1993) advocate the use of "sensitivity" indices and main effect (and interaction) plots to assess the sensitivity of the output $y(\cdot)$ to individual variables or combinations of variables (see Saltelli et al. (2000) for a detailed summary). Both the indices and plots are based on an ANOVA-type decomposition of the function $y(\boldsymbol{x}) = y(x_1, x_2, \ldots, x_d)$ having d input variables into components of increasing dimension. For simplicity, suppose our input region is the d-dimensional unit cube $[0,1]^d$.

Let

$$y_0 = \int_{[0,1]^d} y(x_1, \ldots, x_d)\, dx_1 \cdots dx_d$$

be the overall mean of $y(\boldsymbol{x})$. Sobol´ (1993) shows that there is a unique decomposition

$$\begin{aligned} y(x_1, \ldots, x_d) = y_0 + \sum_{i=1}^{d} y_i(x_i) + \\ \sum_{1 \le i < j \le d} y_{ij}(x_i, x_j) + \cdots + y_{1,2,\ldots,d}(x_1, \ldots, x_d), \quad (7.1.1) \end{aligned}$$

that satisfies

$$\int_0^1 y_{i_1,\ldots,i_s}(x_{i_1}, \ldots, x_{i_s})\, dx_{i_k} = 0$$

for any $1 \le k \le s$ and has orthogonal components. The latter means that if $(i_1, \ldots, i_s) \ne (j_1, \ldots, j_t)$,

$$\int_{[0,1]^d} y_{i_1,\ldots,i_s}(x_{i_1}, \ldots, x_{i_s}) y_{j_1,\ldots,j_t}(x_{j_1}, \ldots, x_{j_t})\, dx_1 \cdots dx_d = 0.$$

For example,

$$y_i(x_i) = \int_0^1 \cdots \int_0^1 y(x_1, \ldots, x_d)\, d\boldsymbol{x}_{-i} \; - \; y_0$$

is the main effect of input i, and

$$y_{ij}(x_i, x_j) = \int_0^1 \cdots \int_0^1 y(x_1, \ldots, x_d)\, d\boldsymbol{x}_{-(ij)} \;-\; y_0 \;-\; y_i(x_i) \;-\; y_j(x_j),$$

is the interaction effect of inputs i and j. Here $d\boldsymbol{x}_{-i}$ denotes integration over all variables except x_i, and $d\boldsymbol{x}_{-(ij)}$ denotes integration over all variables except x_i and x_j.

A *main effect plot* displays $(x_i, y_i(x_i))$ over the range of the ith variable, for $1 \le i \le d$. An *interaction plot* shows (x_i, x_j) versus $y_{ij}(x_i, x_j)$ over the range of the ith and j^{th} variables, for $1 \le i < j \le d$. The output is most sensitive to that variable (or combination of variables) for which the main effect (interaction) plot shows the greatest variability.

Complementing the visualization methods of the previous paragraph are real-valued variance-based indices obtained from the decomposition (7.1.1). Define the total variance V of $y(x_1, x_2, \ldots, .x_d)$ to be

$$V = \int_{[0,1]^d} y^2(x_1, \ldots, x_d)\, dx_1 \cdots dx_d \;-\; y_0^2.$$

Partial variances are computed from each of the terms in the decomposition as

$$V_{i_1,\ldots,i_s} = \int_0^1 \cdots \int_0^1 y^2_{i_1,\ldots,i_s}(x_{i_1}, \ldots, x_{i_s})\, dx_{i_1} \cdots dx_{i_s}, \tag{7.1.2}$$

for $s = 1, \ldots, d$ and $1 \le i_1 < \cdots < i_s \le d$. If one squares both sides of the decomposition of $y(\cdot)$ and integrates both sides over $[0,1]^d$, one obtains

$$V = \sum_{i=1}^{d} V_i + \sum_{1 \le i < j \le d} V_{ij} + \cdots + V_{1,2,\ldots,d}$$

because of the orthogonality properties of the terms in the decomposition.

For any $s = 1, \ldots, d$ and $1 \le i_1 < \cdots < i_s \le d$, we define sensitivity measures $S_{i_1,\ldots,i_s}$ to be

$$S_{i_1,\ldots,i_s} = \frac{V_{i_1,\ldots,i_s}}{V}.$$

S_i is called the *first-order sensitivity index* for input x_i. It measures the main effect of x_i on the output, i.e., the proportion of the variation V that is due to input x_i. For $i < j$, S_{ij} is called the *second-order sensitivity index*. S_{ij} measures the interaction effect due to inputs i and j, i.e., the proportion of V that is due to inputs x_i and x_j that cannot be explained by their main effects. Higher-order sensitivity indices are defined analogously. By construction, the sensitivity indices satisfy

$$\sum_{i=1}^{d} S_i + \sum_{1 \le i < j \le d} S_{ij} + \cdots + S_{1,2,\ldots,d} = 1.$$

- If the predicted output is not sensitive to an input variable (as measured by the total sensitivity), that variable can be fixed in subsequent modeling of the output.

7.2 Model Validation

7.2.1 Introduction

This section introduces the important problem of computer model validation, i.e., determining if the computer model represents the reality that the code is meant to describe. The fundamental assumption of this section is that field data, produced by physical experiments, provide gold standard information about the true relationship between a set of inputs and an output (or multiple outputs). Both Bayarri, Berger, Higdon, Kennedy, Kottas, Paulo, Sacks, Cafeo, Cavendish, Lin and Tu (2002) and Hills and Trucano (1999) present several specific case studies of model validation. As motivating examples, we describe two studies from Bayarri et al. (2002). In each case, it is instructive for readers to speculate how they would conduct a validation experiment.

One of these studies concerned the validation of a computer model that was used to determine the yield stress and other material properties of a spot weld. Spot welds are formed between two pieces of sheet metal by tightly pressing the pieces between two electrodes and applying a high voltage at this spot to melt the metal pieces and form a bond. Important factors in such a process include the load/pressure at the weld site, the voltage, and the electrical resistance between the electrodes. A second study concerned a computer model that generated the velocity $\times$ time deceleration profile at critical positions in a motor vehicle undergoing a front-end crash into a fixed barrier; one such position is the passenger compartment. Among other factors, the severity of the crash is determined by the initial vehicle velocity and the angle at which the vehicle hits the crash barrier. While it is clearly substantially more expensive to crash automobiles into barriers than it is to weld test pieces of sheet metal, in both cases it is possible to gather field data for validation of the code.

In some cases it may not be possible to conduct a physical experiment that uses the input/output mechanism that the computer model is meant to describe. However, it may be practical to conduct an experiment that tests a *related* mechanism. Such a "quasi" experiment can permit one to perform a limited validation of the computer model. To illustrate such a situation, recall Chang et al. (2001), who used a computer model to design a hip prosthesis. Their computer code produced the stresses and strains along the prosthesis $\times$ bone interface. An experiment could be run that tests the output of this computer model only by implanting prostheses that have been instrumented with (numerous) strain gauges into a group of patients having bone characteristics and gait patterns similar to those in the target population for the prosthesis. Clearly, there would be ethical difficulties in conducting such an experiment.

Instead, Chang et al. (2001) conducted a partial validation study based on the following physical experiment. A set of "sawbones" (synthetic material formed in the shape of the human femur) was implanted with hip prostheses whose (b, d) dimensions spanned the same range studied in the computer experiment (see Figure 1.2). The material properties of the sawbones mimicked those found in the human population at risk for hip replacement. Each prosthesis contained a set of strain gauges that measured the stresses at multiple (internal) sites along the prosthesis $\times$ bone interface during cyclic loading. In addition to the bone elasticities and prosthesis dimensions, the load and friction were varied in this experiment. Validation was performed by *identifying the designs that were judged best* based on this "quasi-clinical" experiment with those judged best based on the computer experiment. Sacks, Rouphail, Park and Thakuriah (2000) give another example in which it was impossible to conduct a physical experiment to validate a computer model. In their case, the computer code CORSIM was used to model traffic flow and again, a limited type of quasi-validation was possible.

Finally there are some settings where it is, indeed, impossible to run either a physical experiment or quasi-experiment to validate a computer model. As mentioned in Chapter 1, the fact that a physical experiment is impossible is one reason that motivates some researchers to use computer experiments. Models that describe the evolution of wildfires and weather prediction are of this type. If one is unfortunate enough to be in this situation, *caveat emptor* applies; the only way of ensuring a computer model is able to predict reality is by a validation experiment. The same situation occurs in many other settings. The impossibility of using a physical experiment to study the role of certain behaviors on the incidence of various types of cancer has led to the development of observational studies and other data collection and data analysis methods. Similarly Berk et al. (2002) discuss the use of alternate studies to provide some level of computer model validation.

The remainder of this section restricts attention to cases where it is possible to conduct a validation experiment. We outline the statistical model used by Bayarri et al. (2002) to carry out a Bayesian validation analysis. The elements of this model are also present in the work of other authors who have discussed this problem. For example, Easterling (2001) presents a frequentist analysis of validation experiment data based on a model having similar characteristics. Before turning to the model, we describe two problems that are closely related to model validation—calibration of computer models and combining the results of physical and computer experiments.

7.2.2 Related Problems

Model calibration is the problem of choosing the model parameters of a computer code so that field data are well-approximated by the computer model. Recall that model parameters are called model variables and denoted by $\boldsymbol{x}_m$ in Section 2.1. A key element of any such analysis is the choice of a "distance" that is used to measure the discrepancy between the computer and field data (see Trucano, Pilch and Oberkampf (2002) and Pilch, Trucano, Moya, Froehlich, Hodges and Peercy (2000) for discussions of metrics). The problem of model calibration has been addressed both from a frequentist viewpoint (Park (1991)) and from the Bayesian viewpoint (Craig et al. (2001), Kennedy and O'Hagan (2001) and the references therein).

The idea of *combining the results* of a physical experiment with the results of a computer experiment is attractive. Research on this problem is in its infancy. One report that addresses it from a Bayesian viewpoint is Reese et al. (2000) whose formulation even allows data from uncalibrated computer experiments.

Model validation is concerned with assessing whether data from a computer model approximates field data to an acceptable degree of accuracy. The approximation can be unacceptable for many reasons. There can be inadequacies in the mathematical modeling of the physical phenomenon of interest (simplification in its critical physics or biology; even if the physics or biology is correct, the detailed choices of model parameters—for example, metabolism rates—can be in error). There can be problems with the numerical algorithm(s) used to implement the mathematical model.

Although the phrases sound similar, model "validation" should be distinguished from computer model "verification." In contrast to validation, which determines whether a computer code describes reality, *model verification* is the process of confirming that solutions produced by a numerical algorithm approximate those of a given theoretical mathematical model to some desired level of accuracy. As an example, showing that a finite difference solution accurately solves a differential equation model would verify this mathematical model. It is important that the input domain, the region of the input space over which the code is to provide evaluations, be specified. Every numerical method has limitations (deviations of the output from "true," analytically computed output values that are due to hardware or software can be thought of as "numerical noise"). The critical issue is whether the code at hand provides output sufficiently near analytically computed outputs for inputs over the desired domain. Comparisons of the output with analytic solutions are but one method of model verification. If the computer model output is not verified, we cannot be sure what mathematical model is being tested. A model must be verified before the validation process proceeds. See Roache (1998), Hills and Trucano (1999), and Pilch et al. (2000) for detailed discussions of model verification.

7.2.3 Assessing Validation

The conceptual models used as the statistical basis for analyzing problems of model validation, calibration, and combining data from multiple experimental modes share several guiding principles.

1. There is an underlying true input-output function that describes reality.

2. Computer output is a biased version of reality. Some reasons for this bias are: the computer model contains calibration parameters that are *incorrectly specified*; there are variables that the computer model *omits* but which must be specified in the true input-output function; and the computer model uses either incorrect or overly simplified physics or biology to describe the true input-output relation.

3. Physical experiments are noisy versions of reality. While some of the deviations from reality can be due to traditional measurement error sources, at least part of the noise is often caused by latent variables that are present in the physical experiment but which are not accounted for by the computer model.

The statistical models used by Reese et al. (2000), Craig et al. (2001), Easterling (2001), Kennedy and O'Hagan (2001), and Bayarri et al. (2002) all contain features of the three principles sketched above.

As a specific example, the model of Bayarri et al. (2002) relating reality, the computer code, and the physical experimental output can be described as follows. Let $\boldsymbol{x}_c$ denote the vector of user-controlled inputs and $y^T(\boldsymbol{x}_c)$ the true output at $\boldsymbol{x}_c$. Among the biases in the computer model that they allow are *unknown* model variables, i.e., the computer code is assumed to be of the form $y^C(\boldsymbol{x}_c, \boldsymbol{x}_m)$, where $\boldsymbol{x}_c$ is the vector of control variables and $\boldsymbol{x}_m$ is the vector of model variables. While the computer model can be run at an arbitrary $\boldsymbol{x}_m$, we denote the unknown model variable used by the true input/output function $y^T(\boldsymbol{x}_c)$ to be $\boldsymbol{x}_m^T$. Bayarri et al. (2002) assume the true output at $\boldsymbol{x}_c$ is related to the computer output by

$$y^T(\boldsymbol{x}_c) = y^C(\boldsymbol{x}_c, \boldsymbol{x}_m^T) + b(\boldsymbol{x}_c), \qquad (7.2.1)$$

where $b(\cdot)$ is an *unknown bias* function that describes the deviations of the computer model from reality. Notice that (7.2.1) implicitly assumes that $b(\boldsymbol{x}_c)$ is the bias when *the calibration parameters are correctly set.* Thus $b(\boldsymbol{x}_c)$ is meant to capture all the computer model inadequacies aside from those caused by the use of an incorrect calibration parameter. If the computer model is run using model variable $\boldsymbol{x}_m \neq \boldsymbol{x}_m^T$, then the bias must be changed to incorporate this additional inadequacy and the more complete notation $b(\boldsymbol{x}_c, \boldsymbol{x}_m)$ describes better the bias. As usual, the prediction methods discussed in Chapter 3 can be used to approximate $y^C(\cdot, \cdot)$. If

$(\boldsymbol{x}_{c,i}, \boldsymbol{x}_{m,i})$, $1 \le i \le n_c$ denote the training data for the computer model, then the EBLUPs of Section 3.3 are well-suited for this purpose.

When run at input $\boldsymbol{x}_{c,i}$, the output of the physical experiment is modeled as

$$y^F(\boldsymbol{x}_{c,i}) = y^T(\boldsymbol{x}_{c,i}) + \epsilon_i, \qquad (7.2.2)$$

where $y^F(\cdot)$ denotes the field data and the $\{\epsilon_i\}$ are measurement errors. Suppose field data are observed at n_f input sites. Then an extension of the EBLUP that incorporates a "nugget" effect can be used to predict $y^T(\cdot)$ based on these n_f observations. Model (7.2.2) is the usual quantification of the fact that experimental data are noisy versions of reality.

Often it may be useful to acknowledge that the physical experiment may be imperfect by allowing a bias term in the model,

$$y^F(\boldsymbol{x}_{c,i}) = y^T(\boldsymbol{x}_{c,i}) + b^F(\boldsymbol{x}_{c,i}) + \epsilon_i \ .$$

The bias function $b^F(\cdot)$ need not be estimable depending on the experimental designs used to generate the computer and field data. In either case, the $\{\epsilon_i\}$ are assumed to be $N(0, \sigma_F^2)$ and either independent or possibly correlated, depending on the experimental design.

The design problem is concerned with the choice of the computer and physical experimental inputs. The choice of a design is dependent on the goal of the validation analysis. For example, in model (7.2.1)-(7.2.2), some authors view the goal as that of using the computer and physical data to test the hypothesis that the bias function $b(\cdot)$ is zero, at least over a target input range. Other authors regard the problem as that of forming predictive bands (frequentist or posterior, Bayesian) on the bias function. A third possibility of assessing validation is to assure that the bias is sufficiently small that, if inputs are ordered by their $y^T(\boldsymbol{x}_c)$ values, then the same input ordering holds for $y^C(\boldsymbol{x}_c, \boldsymbol{x}_m^T)$. The statistical community has not coalesced ideas about the proper formulation of the analysis (and hence how to design such experiments); the entire area of model validation is one of active research.

Appendix B
Mathematical Facts

B.1 The Multivariate Normal Distribution

There are several equivalent of ways of defining the multivariate normal distribution. Because we mention both degenerate ("singular") and nondegenerate ("nonsingular") multivariate normal distributions, we will define this distribution by the standard device of describing it indirectly as the distribution that arises by forming a certain function, affine combinations, of independent and identically distributed standard normal random variables.

Definition Suppose $\boldsymbol{Z} = (Z_1, \ldots, Z_r)$ consists of independent and identically distributed $N(0,1)$ random variables, $\boldsymbol{L}$ is an $m \times r$ real matrix, and μ is $m \times 1$ real vector. Then

$$\boldsymbol{W} = (W_1, \ldots, W_m) = \boldsymbol{L}\boldsymbol{Z} + \boldsymbol{\mu}$$

is said to have the *multivariate normal distribution* (associated with $\boldsymbol{\mu}$, $\boldsymbol{L}$).

It is straightforward to compute the mean vector of $(W_1, \ldots, W_m)$ and the matrix of the variances and covariances of the $(W_1, \ldots, W_m)$ in terms of $(\boldsymbol{\mu}, \boldsymbol{L})$ as

$$\boldsymbol{\mu} = E\{\boldsymbol{W}\} \text{ and } \mathrm{Cov}\{\boldsymbol{W}\} = E\{(\boldsymbol{W} - \boldsymbol{\mu})(\boldsymbol{W} - \boldsymbol{\mu})^\top\} = \boldsymbol{L}\boldsymbol{L}^\top.$$

As an example, suppose Z_1 and Z_2 are independent $N(0,1)$ and

$$\boldsymbol{W} = \begin{pmatrix} W_1 \\ W_2 \end{pmatrix} = \boldsymbol{LZ} + \boldsymbol{\mu} = \begin{pmatrix} 3 & 0 \\ 5 & 0 \end{pmatrix} \begin{pmatrix} Z_1 \\ Z_2 \end{pmatrix} + \begin{pmatrix} 2 \\ 0 \end{pmatrix}.$$

By construction, $\boldsymbol{W}$ has a multivariate normal distribution. Notice that both W_1 and W_2 can be expressed in terms of the other. In contrast,

$$\begin{pmatrix} W_1 \\ W_2 \end{pmatrix} = \begin{pmatrix} 3 & 2 \\ 0 & 5 \end{pmatrix} \begin{pmatrix} Z_1 \\ Z_2 \end{pmatrix} + \begin{pmatrix} 0 \\ 0 \end{pmatrix}$$

also has a multivariate normal distribution. In the second case we have defined the same number of linearly independent W_is as independent Z_is.

This example illustrates the fundamental dichotomy in multivariate normal distributions. A multivariate normal distribution is *nonsingular* (nondegenerate) if the rows of $\boldsymbol{L}$ are linearly independent, i.e, $\text{rank}(\boldsymbol{L}) = m$, and it is *singular* (degenerate) if the rows of $\boldsymbol{L}$ are linearly dependent, i.e, $\text{rank}(\boldsymbol{L}) < m$.

Suppose $\boldsymbol{W}$ has the nonsingular multivariate normal distribution defined by $\boldsymbol{\mu} \in \mathbb{R}^m$ and $m \times r$ matrix $\boldsymbol{L}$ having rank m. Let

$$\boldsymbol{\Sigma} = \boldsymbol{L}\boldsymbol{L}^\top$$

denote the covariance matrix of $\boldsymbol{W}$. Notice that $\boldsymbol{\Sigma}$ must be symmetric and positive definite (the latter follows because if $\|\cdot\|_2$ denotes Euclidean norm and $\boldsymbol{z} \neq \boldsymbol{0}$, then $\boldsymbol{z}^\top \boldsymbol{\Sigma} \boldsymbol{z} = \boldsymbol{z}^\top \boldsymbol{L}\boldsymbol{L}^\top \boldsymbol{z} = \|\boldsymbol{L}^\top \boldsymbol{z}\|_2^2 > 0$ because $\text{rank}(\boldsymbol{L}) = m$). In this case it can be shown that $\boldsymbol{W} = (W_1, \ldots, W_m)$ has density

$$f(\boldsymbol{w}) = \frac{1}{(2\pi)^{m/2}(\det(\boldsymbol{\Sigma}))^{1/2}} \exp\left\{ -\frac{1}{2}(\boldsymbol{w} - \boldsymbol{\mu})^\top \boldsymbol{\Sigma}^{-1} (\boldsymbol{w} - \boldsymbol{\mu}) \right\} \quad \text{(B.1.1)}$$

over $\boldsymbol{w} \in \mathbb{R}^m$. We denote the fact that $\boldsymbol{W}$ has the nonsingular multivariate normal distribution (B.1.1) by $\boldsymbol{W} \sim N_m(\boldsymbol{\mu}, \boldsymbol{\Sigma})$. There are numerous algorithms for computing various quantiles associated with multivariate normal distributions. We note, in particular, Dunnett (1989) who provides a FORTRAN 77 program for computing equicoordinate percentage points of multivariate normal distributions having product correlation structure (see also Odeh, Davenport and Pearson (1988)).

Now suppose $\boldsymbol{W}$ has the singular multivariate normal distribution defined by $\boldsymbol{\mu} \in \mathbb{R}^m$ and $m \times r$ matrix $\boldsymbol{L}$ where $\text{rank}(\boldsymbol{L}) = q < m$. Then $m - q$ rows of $\boldsymbol{L}$ can be expressed as linear combinations of the remaining q rows of $\boldsymbol{L}$ and the corresponding $m - q$ components of $\boldsymbol{W} - \boldsymbol{\mu}$ can be expressed as (the same) linear combinations of the remaining q components of $\boldsymbol{W} - \boldsymbol{\mu}$. Thus, in this case, the support of $\boldsymbol{W}$ is on a hyperplane in a lower dimensional subspace of $\mathbb{R}^m$. Furthermore, the q components of $\boldsymbol{W}$ used to express the remaining variables have a nonsingular multivariate normal distribution with density on $\mathbb{R}^q$.

To illustrate the singular case, consider the toy example above. Marginally, both W_1 and W_2 have proper normal distributions with $W_1 = 3Z_1 + 2 \sim N(2,9)$ and $W_2 = 5Z_1 \sim N(0,25)$. This shows that choice of the q variables that have the proper normal distribution is nonunique. Given either W_1 or W_2, the other can be expressed in terms of the first. For example, given W_1, $W_2 = 5(W_1 - 2)/3$ with probability one or given W_2, $W_1 = 2 + 3W_2/5$ with probability one. In the second part of the example, W_1 and W_2 have a nonsingular bivariate normal distribution.

In the text we make use of the following integration formula, which is an application of the fact that (B.1.1) is a density function.

Lemma B.1.1 For any $n \times 1$ vector $\boldsymbol{v}$ and any $n \times n$ symmetric, positive definite matrix $\boldsymbol{A}$,

$$\int_{I\!R^n} \exp\left\{-\frac{1}{2}\boldsymbol{w}^\top \boldsymbol{A}^{-1}\boldsymbol{w} + \boldsymbol{v}^\top \boldsymbol{w}\right\} d\boldsymbol{w} = (2\pi)^{n/2}(\det(\boldsymbol{A}))^{1/2} \exp\left\{\frac{1}{2}\boldsymbol{v}^\top \boldsymbol{A}\boldsymbol{v}\right\}.$$

To prove this formula consider the $N_n(\boldsymbol{\mu}, \boldsymbol{\Sigma})$ multivariate normal density with covariance matrix $\boldsymbol{\Sigma} = \boldsymbol{A}$ and mean $\boldsymbol{\mu} = \boldsymbol{\Sigma}\boldsymbol{v}$. Then

$$\begin{aligned}(2\pi)^{n/2}(\det(\boldsymbol{\Sigma}))^{1/2} &= \int_{I\!R^n} \exp\left\{-\frac{1}{2}(\boldsymbol{w}-\boldsymbol{\mu})^\top \boldsymbol{\Sigma}^{-1}(\boldsymbol{w}-\boldsymbol{\mu})\right\} d\boldsymbol{w} \\ &= \int_{I\!R^n} \exp\left\{-\frac{1}{2}\boldsymbol{w}^\top \boldsymbol{\Sigma}^{-1}\boldsymbol{w} + \boldsymbol{\mu}^\top \boldsymbol{\Sigma}^{-1}\boldsymbol{w} - \frac{1}{2}\boldsymbol{\mu}^\top \boldsymbol{\Sigma}^{-1}\boldsymbol{\mu}\right\} d\boldsymbol{w}.\end{aligned}$$

Substituting for $\boldsymbol{\Sigma}$ and $\boldsymbol{\mu}$ and rearranging terms gives the result. □

Perhaps more usefully, we can interpret the proof of Lemma B.1.1 as stating that if $\boldsymbol{W}$ has density $f(\boldsymbol{w})$, for which

$$f(\boldsymbol{w}) \propto \exp\left\{-\frac{1}{2}\boldsymbol{w}^\top \boldsymbol{A}^{-1}\boldsymbol{w} + \boldsymbol{v}^\top \boldsymbol{w}\right\}, \text{ then } \boldsymbol{W} \sim N_n\left[\boldsymbol{A}\boldsymbol{v}, \boldsymbol{A}\right]. \quad \text{(B.1.2)}$$

We also require the following result concerning the conditional distribution of a set of components of the multivariate normal distribution given the remaining ones.

Lemma B.1.2 (Conditional distribution of the multivariate normal) Suppose that

$$\begin{pmatrix} W_1 \\ W_2 \end{pmatrix} = N_{m+n}\left[\begin{pmatrix} \boldsymbol{\mu}_1 \\ \boldsymbol{\mu}_2 \end{pmatrix}, \begin{pmatrix} \boldsymbol{\Sigma}_{1,1} & \boldsymbol{\Sigma}_{1,2} \\ \boldsymbol{\Sigma}_{2,1} & \boldsymbol{\Sigma}_{2,2} \end{pmatrix}\right]$$

where $\boldsymbol{\mu}_1$ is $m \times 1$, $\boldsymbol{\mu}_2$ is $n \times 1$, $\boldsymbol{\Sigma}_{1,1}$ is $m \times m$, $\boldsymbol{\Sigma}_{1,2} = \boldsymbol{\Sigma}_{2,1}^\top$ is $m \times n$, and $\boldsymbol{\Sigma}_{2,2}$ is $n \times n$. Then the conditional distribution of $W_1|W_2$ is

$$N_m\left[\boldsymbol{\mu}_1 + \boldsymbol{\Sigma}_{1,2}\boldsymbol{\Sigma}_{2,2}^{-1}(W_2 - \boldsymbol{\mu}_2), \boldsymbol{\Sigma}_{1,1} - \boldsymbol{\Sigma}_{1,2}\boldsymbol{\Sigma}_{2,2}^{-1}\boldsymbol{\Sigma}_{2,1}\right].$$

B.2 The Non-Central Student t Distribution

This appendix defines the univariate Student t and multivariate Student t distributions. Throughout, suppose that $\boldsymbol{\mu} \in \mathbb{R}^m$ and $\boldsymbol{\Sigma}$ is a positive definite matrix.

Definition The random vector $\boldsymbol{W} = (W_1, \ldots, W_m)$ with joint probability density

$$f(\boldsymbol{w}) = \frac{\Gamma((\nu+m)/2)}{(\det(\boldsymbol{\Sigma}))^{1/2}(\nu\pi)^{m/2}\Gamma(\nu/2)} \left(1 + \frac{1}{\nu}(\boldsymbol{w}-\boldsymbol{\mu})^\top \boldsymbol{\Sigma}^{-1}(\boldsymbol{w}-\boldsymbol{\mu})\right)^{-(\nu+m)/2} \tag{B.2.1}$$

over $\boldsymbol{w} \in \mathbb{R}^m$ is said to have the *nonsingular multivariate t distribution* with ν degrees of freedom, location parameter $\boldsymbol{\mu}$, and scale matrix $\boldsymbol{\Sigma}$.

We denote the multivariate t distribution (B.2.1) by $\boldsymbol{W} \sim T_m(\nu, \boldsymbol{\mu}, \boldsymbol{\Sigma})$. The $T_m(\nu, \boldsymbol{\mu}, \boldsymbol{\Sigma})$ distribution has mean vector $\boldsymbol{\mu}$ provided $\nu > 1$ and has covariance matrix $\nu\boldsymbol{\Sigma}/(\nu-2)$ provided $\nu > 2$. The "usual" univariate t and multivariate t distributions are the special cases of (B.2.1) corresponding to

- $T_1(\nu, 0, 1)$ and
- $T_m(\nu, \mathbf{0}, \boldsymbol{R})$

where $\boldsymbol{R}$ has unit diagonal elements (Tong (1980), Odeh et al. (1988) and Dunnett (1989)).

In particular, if $\boldsymbol{X} \sim N_m(\mathbf{0}, \boldsymbol{R})$, where $\boldsymbol{R}$ is as in the previous paragraph, and $\boldsymbol{X}$ is independent of $V \sim \chi^2_\nu$, then

$$\boldsymbol{W} = (W_1, \ldots, W_m) \equiv \left(\frac{X_1}{\sqrt{V/\nu}}, \ldots, \frac{X_m}{\sqrt{V/\nu}}\right) \sim T_p(\nu, 0, \boldsymbol{R}).$$

Some other important relationships concerning the multivariate t distribution are

- If $\boldsymbol{W} \sim T_p(\nu, \boldsymbol{\mu}, \boldsymbol{\Sigma})$ then $\boldsymbol{W} - \boldsymbol{\mu} \sim T_p(\nu, \mathbf{0}, \boldsymbol{\Sigma})$.
- When $p = 1$, $W \sim T_1(\nu, \mu, \sigma)$ if and only if $\frac{1}{\sigma}(W - \mu) \sim T_1(\nu, 0, 1)$.
- For arbitrary p, $\mu \in \mathbb{R}^p$, and $p \times p$ positive definite $\boldsymbol{\Sigma}$ with diagonal elements σ_i^2 for $1 \le i \le p$, $\boldsymbol{W} \sim T_p(\nu, \boldsymbol{\mu}, \boldsymbol{\Sigma})$ if and only if $\boldsymbol{\Lambda}^{-1}(\boldsymbol{W} - \boldsymbol{\mu}) \sim T_p(\nu, 0, \boldsymbol{R})$ where

$$\boldsymbol{\Lambda} = \boldsymbol{\Lambda}^\top = \text{diag}(\sigma_1, \ldots, \sigma_m).$$

Dunnett (1989) provides an extension of his multivariate normal percentile program to compute equicoordinate percentage points of the multivariate t distribution for the case where the scale matrix has unit variances and a (rank one) product correlation structure.

3. **Matérn I**

$$R(\boldsymbol{h}) = \prod_{i=1}^{d} \frac{1}{\Gamma(\nu)2^{\nu-1}} \left(\frac{2\sqrt{\nu}\,|h_i|}{\theta_i^m} \right)^{\nu} K_{\nu} \left(\frac{2\sqrt{\nu}\,|h_i|}{\theta_i^m} \right) \qquad \text{(C.1.3)}$$

where $\theta_i^m > 0$, $1 \le i \le d$, and $\nu > 0$ are unknown while K_ν is the modified Bessel function of order ν (Stein (1999)). Thus $\boldsymbol{\psi} = (\theta_1^m, \ldots, \theta_d^m, \nu)$.

4. **Isotropic Matérn I**

$$R(\boldsymbol{h}) = \frac{1}{\Gamma(\nu)2^{\nu-1}} \left(\frac{2\sqrt{\nu}\,\|\boldsymbol{h}\|_2}{\theta^m} \right)^{\nu} K_{\nu} \left(\frac{2\sqrt{\nu}\,\|\boldsymbol{h}\|_2}{\theta^m} \right) \qquad \text{(C.1.4)}$$

where $\theta^m > 0$ is unknown so that $\boldsymbol{\psi}$ is the vector (θ^m, ν).

5. **Matérn II**

$$R(\boldsymbol{h}) = \prod_{i=1}^{d} \frac{1}{\Gamma(\nu_i)2^{\nu_i-1}} \left(\frac{2\sqrt{\nu_i}\,|h_i|}{\theta_i^m} \right)^{\nu_i} K_{\nu_i} \left(\frac{2\sqrt{\nu_i}\,|h_i|}{\theta_i^m} \right) \qquad \text{(C.1.5)}$$

where $\theta_i^m > 0$ and $\nu_i > 0$ are unknown for $1 \le i \le d$. In this case, $\boldsymbol{\psi} = (\theta_1^m, \ldots, \theta_d^m, \nu_1, \ldots, \nu_d)$.

As noted in Chapter 2, ν (in the case of the Matérn I family) and each ν_i (in case of the Matérn II family) are smoothness parameters governing the differentiability of sample paths drawn from the process with Matérn correlation function. These three families are (partially) increasing in the sense that the Matérn II family obviously includes the Matérn I family *and*, as $\nu \to \infty$, the ith term in (C.1.3) converges to the ith term in (C.1.1) with $p_i = 2$ and $\theta_i^e = 1/(\theta_i^m)^2$. Thus the power exponential family with quadratic power is a limiting case of the Matérn I family.

This program can obtain estimates of the $\boldsymbol{\psi}$ correlation parameters by maximum likelihood estimation, by restricted maximum likelihood estimation, or by the maximum of the posterior mode (MAP) based on certain priors for $(\boldsymbol{\beta}, \sigma_Z^2, \boldsymbol{\psi})$. The priors that `PErK` allows assume independence of $(\boldsymbol{\beta}, \sigma_Z^2)$ and $\boldsymbol{\psi}$. It allows any of the priors specified in Theorem 4.1.2 for $(\boldsymbol{\beta}, \sigma_Z^2)$. `PErK` assumes that the components of the correlation parameter $\boldsymbol{\psi}$ are independent. Each power p_i (or p) is a mixture of a point mass at $p_i = 2$ and a scaled beta distribution on $(0, 2)$. Each ν_i (or ν) is a scaled beta distribution on $(0, 50)$. All range parameters have gamma distributions.

`PErK` carries out optimizations by first applying the simplex algorithm of Nelder and Mead, followed by a quasi-Newton algorithm using finite-difference gradients.

In addition to being able to predict $y(\boldsymbol{x})$ at arbitrary sites based on training data, `PErK` can also predict linear combinations of $y(\boldsymbol{x})$. This application can be motivated as follows. In some cases, the input vector $\boldsymbol{x}$

consists of two components, $\boldsymbol{x} = (\boldsymbol{x}_c, \boldsymbol{x}_e)$ where $\boldsymbol{x}_c$ denotes a vector of *control* (engineering design) variables and $\boldsymbol{x}_e$ denotes a vector of *environmental* variables. Suppose that the support of the environmental variables is finite on the points $\{\boldsymbol{x}^{sp}_{e,i}\}_{i=1}^{n_e}$ and has associated probabilities (weights) $\{w_i\}_{i=1}^{n_e}$; then PErK can automatically provide predictions of the linear combination

$$\mu(\boldsymbol{x}_c) \equiv \sum_{i=1}^{n_e} w_i \, y(\boldsymbol{x}_c, \boldsymbol{x}^{sp}_{e,i}) \, .$$

$\mu(\boldsymbol{x}_c)$ is the mean of $y(\boldsymbol{x}_c, \boldsymbol{X}_e)$ with respect to the environmental variable distribution.

The next section describes the syntax for the required and optional PErK commands. The third section gives a series of examples illustrating the syntax.

C.2 PErK Job File Options and Output

A. Definition stage of PErK job

Required Commands: The following objects *must* be defined. However, if the user wishes to use the default value for any of these objects, the object need not be explicitly included in the job file.

- `CorrelationFamily = corclass` where `corclass` specifies the parametric family of correlation functions used for data analysis. Currently, `corclass` can assume the value `Matern` or `PowerExponential`.
- `CorrelationType = ctype` where `ctype` specifies the nature of the parameters of the correlation function. If `CorrelationFamily = PowerExponential`, then `ctype = 2` specifies that the product (C.1.1) is to be used while if `ctype = 3`, then the isotropic (C.1.2) is to be used. If `CorrelationFamily = Matern`, `ctype = 1` indicates that the unconstrained Matérn I correlation function (C.1.3) is to be used, `ctype = 2` tells the program to use the Matérn II family (C.1.5) with separate smoothness parameters for each input variable, and `ctype = 3` specifies use of the two-parameter isotropic Matérn I correlation function (C.1.4). **Default**: `CorrelationFamily` = `PowerExponential` when `ctype = 2` and = `Matern` when `ctype = 1`.
- `RandomError = randerr` where `randerr` indicates whether or not a white noise process is included in the model. It takes the value `Yes` or `No`. **Default**: `randerr = No`.
- `CorrelationEstimation = mthd` where `mthd` describes the method of estimating the correlation parameters; `mthd` takes the value `MLE`,

REML, or MAP. The MLE chooses the correlation parameters ψ to maximize the log likelihood $\ell(\boldsymbol{\beta}, \sigma_Z^2, \psi)$ of the training data in (3.3.11) (after expressing $\boldsymbol{\beta}$ and σ_Z^2 as functions of ψ). The REML of the correlation parameters ψ maximizes the log likelihood of a modified version of the training data that is independent of parameters that determine the mean. The MAP option chooses the correlation parameters ψ to maximize the posterior distribution $[\psi | \boldsymbol{Y}^n]$ in (4.1.20).

Use of the MAP method *requires* the user to create additional input for each $[\boldsymbol{\beta}, \sigma_Z^2]$ prior in (1), (2), or (3) in

	$[\sigma_Z^2]$	
$[\boldsymbol{\beta} \mid \sigma_Z^2]$	$c_0 \times \chi_{\nu_0}^{-2}$	$1/\sigma_Z^2$
$N(\boldsymbol{b}_0, \sigma_Z^2 \boldsymbol{V}_0)$	(1)	(2)
1	(3)	(4)

The MAP estimator of ψ that corresponds to the non-informative prior

$$[\boldsymbol{\beta}, \sigma_Z^2] \propto \frac{1}{\sigma_Z^2},$$

is the REML estimator of ψ. The $\boldsymbol{b}_0$ and σ_Z^2 of the proper $N(\boldsymbol{b}_0, \sigma_Z^2 \boldsymbol{V}_0)$ prior of Theorem 4.1.2 are specified by the input objects `PriorBetaMean` and `PriorBetaVariance` (see below). The hyperparameters c_0 and ν_0 of the $c_0 \times \chi_{\nu_0}^{-2}$ prior are specified by the input objects `PriorSigmaZSqC0` and `PriorSigmaZSqV0` (also see below). **Default**: `CorrelationEstimation = REML`.

- `Tries = ntries` where `ntries` specifies the number of attempts made at global maximization of the log likelihood function, restricted log likelihood function or log posterior distribution of the correlation parameters. **Default**: `ntries = 5`.

- `BoundaryTries = bntries` where `bntries` specifies the number of attempts made at maximizing the log likelihood function, restricted log likelihood function or log posterior distribution of the correlation parameters assuming infinitely differentiable realizations of the Gaussian random function. **Default**: `bntries = 1`.

- `LogLikelihoodTolerance = logliktol` where `logliktol` specifies the maximum relative tolerance for termination of an attempt at maximizing the log likelihood function, restricted log likelihood function or log posterior distribution of the correlation parameters using the simplex method. **Default**: `logliktol = 1.e-5`.

- `SimplexTolerance = simptol` where `simptol` specifies the maximum relative tolerance for termination of a single run of the simplex algorithm. **Default**: `simptol = 1.e-10`.

- `LogLikelihoodDifference = loglikdiff` where `loglikdiff` specifies the maximum deviation from the optimum log likelihood function, restricted log likelihood function or log posterior value that will allow selection of an infinitely differentiable field in place of the field chosen by ML, REML, or MAP in the event that it is not infinitely differentiable. **Default**: `loglikdiff = 1.0`.

- `RandomNumberSeed = iseed` where `iseed` specifies a starting seed for initialization of the uniform random number generator. The integer `iseed` must be nonzero. **Default**: `iseed = 100`.

- `MaxFunctionIterations = maxfiter` where `maxfiter` specifies the maximum allowable number of evaluations of the log likelihood function, the restricted log likelihood function, or the log posterior distribution for each invocation of the simplex optimization routine. **Default**: `maxfiter = 200`.

- `MaxSimplexIterations = maxsimpiter` where `maxsimpiter` specifies the maximum number of allowable simplex runs for each optimization attempt. **Default**: `maxsimpiter = 5`.

- `Compile = icp` where `icp` indicates whether or not `PErK` is to be compiled upon invocation of a `PErK` job. It takes the value `Yes` or `No`. **Default**: `icp = Yes`.

- `Executable = exec` where `exec` contains the desired name of the executable file. **Default**: `exec = perk.run`.

Optional Commands:

- `CrossValidate` This command is used if the cross-validated predictions of the response and corresponding standard errors are to be computed at each training input $\boldsymbol{x}_i^{tr}$, $i = 1, \ldots, n$. Cross-validated predictions of the response and corresponding standard errors are computed at $\boldsymbol{x}_i^{tr}$ by removing $(\boldsymbol{x}_i^{tr}, y(\boldsymbol{x}_i^{tr}))$ from the training data set and calculating the `PErK` prediction and its standard error based on the reduced $(n-1)$-point input design. To speed computations, the estimates of the correlation parameters and process variance(s) from the full design are used to perform these computations.

B. Input stage of `PErK` job

Required Commands: The following objects *must* be defined. Objects for which a default is indicated need only be included if the default must be superseded.

which they will be obtained. This command is issued only if predictions are required.

- EVSupport < EVpred where EVpred contains the support points in the distribution of the environmental variables.

- EVWeights < wt where wt contains the probabilities in the distribution of the environmental variables corresponding to the support points in EVpred.

- YTrue < ytrue where ytrue contains the values of the response or objective function computed at each prediction site in xpred. This command is issued only if this information is available.

C. Output stage of PErK job

Required Commands: None, but users will want to output one or more of the following files.

Optional Commands: The following objects can optionally be defined.

- Summary > outsummary where outsummary contains summary information about the log likelihood, restricted log likelihood or log posterior maximization, cross-validation, and prediction components of a PErK job. The range of contents of the file outsummary will be demonstrated in the examples.

- RegressionModel > outbeta where outbeta contains estimates of the linear model parameters. Note that these parameter estimates assume that each input variable has been scaled to the unit interval.

- StochasticProcessModel > outcorpar where outcorpar contains the estimates of the correlation parameters. Note that the correlation range parameter estimates assume that each input variable has been scaled to the unit interval. This command need not be issued if values for the correlation parameters are input via the command StochasticProcessModel < incorpar.

- UnformattedSPM > uoutcorpar where uoutcorpar contains the unformatted estimates of the correlation parameters. These are useful when extremely accurate estimates are needed by another program.

- LogLikelihoodSummary > outlogliksum where outlogliksum contains a listing of the log likelihoods, restricted log likelihoods or log posterior values at each exit of the simplex optimization routine for *each* attempt at finding the estimates of the correlation parameters. The correlation parameter estimates upon exiting the simplex and quasi-Newton algorithms are also provided. **Default**: If this object is not specified, then the file `loglik.summary` is created with this information if the job requires optimization of the log likelihood function, restricted log likelihood function, or log posterior distribution.

- `CrossValidations > outcv` where `outcv` contains the cross-validated `PErK` predictions of the response at each input design point. The cross-validated standard errors of prediction are also provided. (See Section 7.1.)

- `Predictions > outpred` where `outpred` contains the `PErK` predictions of the response or objective function corresponding to the prediction sites in `xpred`. The standard errors of prediction are also provided. (See Subsection 3.3.2.)

C.3 Examples

The following examples demonstrate many possible uses of `PErK`. The responses for these examples are based on the *Branin function.* The Branin function is the real-valued function of two variables

$$y_{\mathcal{B}}(x_1, x_2) = \left(x_2 - \frac{5.1}{4\pi^2}\, x_1^2 + \frac{5}{\pi}\, x_1 - 6 \right)^2 + 10 \left(1 - \frac{1}{8\pi} \right) \cos(x_1) + 10$$

on the domain $[-5, 10] \times [0, 15]$ in $\mathbb{R}^2$ which is shown in Figure C.1. For Examples C.1 to C.8, the input function is

$$y(x_1, x_2) = y_{\mathcal{B}}\,(15 \times x_1 - 5, 15 \times x_2) \tag{C.3.1}$$

where $(x_1, x_2) \in [0, 1]^2$ so that the `Ranges` statement is not required.

Example C.1 We calculate the REML estimates of the correlation parameters and corresponding parameters of the linear model

$$Y(\boldsymbol{x}) = \beta_0 + Z(\boldsymbol{x}) \tag{C.3.2}$$

under the assumption that the correlation function of $Z(\cdot)$ is of Matérn I form. Cross-validated predictions and cross-validated standard errors will also be calculated at each input design site. The input sites are located in the file `lhs.21`:

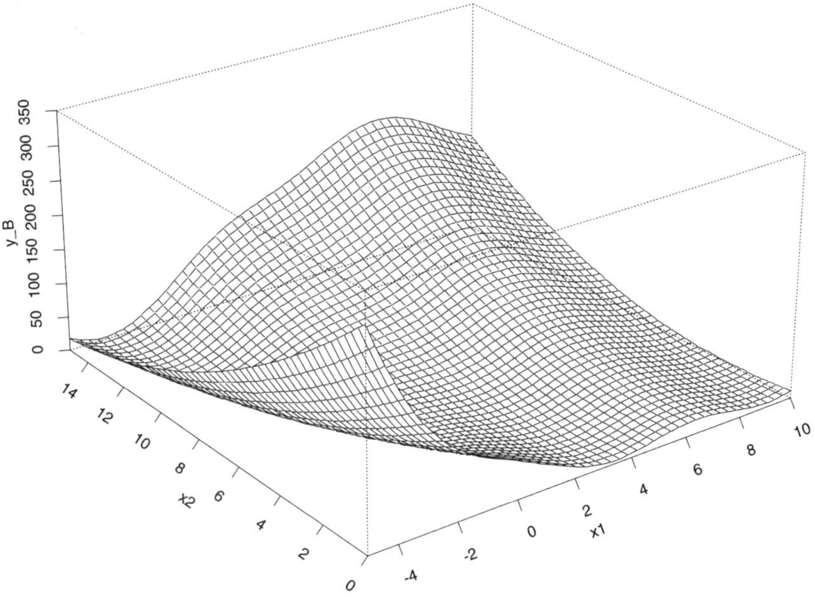

FIGURE C.1. The Branin function on $[-5, 10] \times [0, 15]$

lhs.21

0.83333333	0.40476193
0.40476193	0.26190473
0.97619047	0.54761907
0.64285713	0.30952380
0.50000000	0.97619047
0.11904760	0.16666667
0.54761907	0.02380953
0.02380953	0.45238093
0.07142860	0.83333333
0.73809527	0.11904760
0.88095240	0.73809527
0.78571427	0.92857140
0.30952380	0.07142860
0.21428573	0.35714287
0.35714287	0.69047620
0.92857140	0.21428573
0.16666667	0.64285713
0.69047620	0.59523807
0.59523807	0.78571427
0.26190473	0.88095240
0.45238093	0.50000000

This design is a maximin distance design within the class of Latin Hypercube designs produced by the software Algorithms for the Construction of Experimental Designs (ACED, Welch (1985)). The x_1 values are in the first column of `lhs.21` and the x_2 values are in the second column. The associated values of the response variable are located in the file `y.21`:

y.21

```
35.80951
14.86287
31.41880
19.87899
141.88566
99.43335
3.88973
97.47380
6.27060
19.85914
95.50587
181.74214
49.39445
23.13762
43.09524
2.82392
3.61474
75.79100
104.11175
43.33586
23.39797
```

Because the regression portion of the stochastic process model (C.3.2) has only a constant term, the file **`reg.constant`** is as follows:

reg.constant

```
0
```

The fitting is accomplished by the following `PErK` job file. The *definition*, *input*, and *output* stages of a `PErK` job are indicated by comments in the job file.

PErK job for Example C.1

```
# Definition stage

CorrelationFamily = Matern
CorrelationType = 1 # Specifies Matern I class
RandomError = No
Tries = 5
LogLikelihoodTolerance = 1.e-6
RandomNumberSeed = 300
CrossValidate
CorrelationEstimation = REML

# Input stage

X < lhs.21
Y < y.21
RegressionModel < reg.constant

# Output stage

Summary > C.1.summary
RegressionModel > C.1.beta
StochasticProcessModel > C.1.corpar
CrossValidations > C.1.cv
UnformattedSPM > C.1.unf.corpar
```

The summary information from the fit produced by `PErK` is in the file `C.1.summary`.

C.1.summary

```
Number of Design Sites = 21
Correlation Family = Matern I
Random Error = No
Stochastic Process Variance = 90633600.37871
Restricted Log Likelihood = -56.31388
Number of Restricted Log Likelihood Evaluations = 4179

Condition Number of Cholesky Square Root
of Correlation Matrix = 66155.89301

Cross-Validation RMSPE = 1.32288
Cross-Validation MAD = 3.95518
Case = 6
```

Large values of the condition number, say greater than 10^6, indicate a loss of precision in the parameter estimates due to rounding.

The estimate of the regression mean β_0 is given in `C.1.beta`.

C.1.beta

```
The estimates of the linear model parameters are:

Parameter Beta

1         8903.83812
```

The estimates of the correlation parameters are given in `C.1.corpar`:

C.1.corpar

```
Correlation Family = Matern I

REML estimates of the correlation range parameters are:

Case Range

1         2.44567
2         15.05353

The REML estimate of the correlation smoothness parameter
is:
3.49641
```

The order in which the estimates of the correlation range parameters are presented corresponds to the order of the input variable columns in the file `lhs.21`.

The correlation parameters are output a second time, in an unformatted manner in the file `C.1.unf.corpar`; in this form they can be used again to eliminate the running of the likelihood (restricted likelihood, posterior distribution) optimization. We will illustrate the use of `C.1.unf.corpar` in Example C.7.

C.1.unf.corpar

```
2.4456673998473
15.053534382697
3.4964124536381
```

The first two values will be recognized as the range parameters θ_1^m and θ_2^m in the product Matérn correlation function and the corresponding smoothness parameter is listed last. The `UnformattedSPM` file always contains the set of range parameters followed by the set of smoothness parameters. In

C.2.summary

```
Number of Design Sites = 21
Correlation Family = Matern I
Random Error = No
Stochastic Process Variance = 3182847.61147
Log Likelihood = -62.79702
Number of Log Likelihood Evaluations = 889

Condition Number of Cholesky Square Root
of Correlation Matrix = 6379.05786

Cross-Validation RMSPE = 1.9653
Cross-Validation MAD = 5.20335
Case = 8
```

The estimates of the regression mean and the correlation parameters are given in `C.2.beta` and `C.2.corpar`, respectively.

C.2.beta

```
The estimates of the linear model parameters are:

Parameter Beta

1         2031.63450
```

C.2.corpar

```
Correlation Family = Matern I

MLEs of the correlation range parameters are:

Case Range
1         1.39734
2         6.79094

The MLE of the correlation smoothness parameter is:

3.68699
```

The cross-validated predictions and standard errors of prediction at each input variable site are contained in the file `C.2.cv`:

C.2.cv

```
The Cross-Validation Prediction and Standard Error are:

Case Prediction Standard Error

1          36.505298          1.9898639
2          15.213107          0.8803965
3          29.593050          6.5124128
4          19.119322          1.0356140
5          142.955323          2.3071634
6          96.160093          5.2076946
7          5.411914          1.9688547
8          92.270455          6.6431770
9          9.064924          3.1291044
10          20.096820          2.2147482
11          97.406638          2.4030695
12          178.380573          2.6202900
13          48.407967          1.2839084
14          24.293689          1.0110793
15          43.316083          1.3434502
16          0.078098          4.7059834
17          2.945586          1.0361354
18          76.178488          1.0940489
19          104.508191          1.1197141
20          42.126650          2.2451446
21          22.722972          0.9735322
```

Finally, the MLEs of the correlation parameters for this model are saved in the unformatted form for later use in Example C.8 as the file `C.2.unf.corpar`. ■

Example C.3 We alter Example C.1 by fitting the stochastic process plus measurement error model

$$Y(\boldsymbol{x}) = \beta_0 + Z(\boldsymbol{x}) + \epsilon(\boldsymbol{x}) \tag{C.3.4}$$

to "noisy" training data. Here $Z(\cdot)$ is a zero mean Gaussian process with Matérn I correlation function and $\epsilon(\cdot)$ is zero mean white noise with variance σ_ϵ^2. REML estimates are to be computed for the correlation parameters.

Repeated observations will be taken at some of the input sites to provide "pure error" estimates of σ_ϵ^2 (although this is not a necessary requirement to estimate σ_ϵ^2, assuming the model is true). We add one repeated observation at the second input of `lhs.21` and three repeated observations at the fourteenth input of `lhs.21`. The amended design is placed in the file `lhs.25`. The corresponding responses are located in the file `y.25`:

y.25

```
35.80951
14.06262
14.37753
31.41880
19.87899
141.88566
99.43335
3.88973
97.47380
6.27060
19.85914
95.50587
181.74214
49.39445
23.15795
25.34863
25.50725
22.80699
43.09524
2.82392
3.61474
75.79100
104.11175
43.33586
23.39797
```

The fitting, based on maximum likelihood estimation of correlation parameters, is accomplished using the following `PErK` job file.

PErK job for Example C.3

```
# Definition stage

CorrelationFamily = Matern
CorrelationType = 1
RandomError = Yes
Tries = 5
LogLikelihoodTolerance = 1.e-3
SimplexTolerance = 1.e-3
RandomNumberSeed = 300
CorrelationEstimation = REML

# Input stage

X < lhs.25
Y < y.25
RegressionModel < reg.constant

# Output stage

Summary > C.3.summary
RegressionModel > C.3.beta
StochasticProcessModel > C.3.corpar
```

The results of running `PErK` with this job file are as follows. The summary information, estimate of the regression parameter, and the estimates of the correlation parameters are listed in `C.3.summary`, `C.3.beta`, and `C.3.corpar`, respectively. ■

C.3.summary

```
Number of Design Sites = 25
Correlation Family = Matern I
Random Error = Yes
Stochastic Process Variance = 238807.02526
Error Variance = 1.20618
Restricted Log Likelihood = -61.76074
Number of Restricted Log Likelihood Evaluations = 4542

Condition Number of Cholesky Square Root
of Correlation Matrix = 990.10077
```

C.3.beta

```
The estimates of the linear model parameters are:

Parameter Beta

1         594.66989
```

C.3.corpar

```
Correlation Family = Matern I

REML estimates of the correlation range parameters are:

Case Range

1         0.72071
2         3.32137

The REML of the correlation smoothness parameter is:

4.89426
```

Example C.4 We adjust Example C.1 to fit the model

$$Y(\boldsymbol{x}) = \beta_0 + \beta_1 x_1 + \beta_2 x_2 + \beta_3 x_1 x_2 + Z(\boldsymbol{x}) \tag{C.3.5}$$

where $Z(\cdot)$ has a power exponential correlation function. We also output the fitted correlation parameters for later use in Example C.6. The mean regression model in (C.3.5) is specified in the file `reg.C.4` as follows:

reg.C.4

```
0
1
2
1*2
```

The following job file is used to obtain the `PErK` fit:

PErK job for Example C.4

```
# Definition stage

CorrelationFamily = PowerExponential
RandomError = No
Tries = 5
LogLikelihoodTolerance = 1.e-2
SimplexTolerance = 1.e-2
RandomNumberSeed = 400
CorrelationEstimation = REML

# Input stage

X < lhs.21
Y < y.21
RegressionModel < reg.C.4

# Output stage

Summary > C.4.summary
RegressionModel > C.4.beta
StochasticProcessModel > C.4.corpar
UnformattedSPM > C.4.unf.corpar
```

The results of running `PErK` with this job file are presented below. The summary information from the fit produced by `PErK` is in `C.4.summary`:

C.4.summary

```
Number of Design Sites = 21
Correlation Family = Power Exponential
Random Error = No
Stochastic Process Variance = 193426331.29334
Restricted Log Likelihood = -40.80125
Number of Restricted Log Likelihood Evaluations = 2368

Condition Number of Cholesky Square Root
of Correlation Matrix = 12131.01917
```

The estimate of the regression mean is given in `C.4.beta`:

C.4.beta

```
The estimates of the linear model parameters are:

Parameter Beta

1         12731.53812
2         -3074.46465
3         -1530.09854
4         535.25063
```

The order in which the parameter estimates are presented corresponds to the order of the terms in `reg.C.4`. The estimates of the correlation parameters are listed in `C.4.corpar`.

C.4.corpar

```
Correlation Family = Power Exponential

REML estimates of the correlation parameters are:

Case Range Power 2 - Power

1         0.18495         1.99993         0.00007
2         0.00752         2.00000         0.00000
```

The order in which the estimates of the correlation parameters are presented corresponds to the order of the input variable columns in the file `lhs.21`.

C.4.unf.corpar

```
0.18494678690057
7.5188141072541e-3
7.1497890547390e-5
0.
```

Lastly, we save the estimates of correlation parameters for later use in Example C.6 as `C.4.unf.corpar`. ■

Example C.5 We revise Example C.1 by fitting the model (C.3.2) assuming the Matérn II class of correlation functions for $Z(\cdot)$ and predicting the response at five new input vectors. The input file of the new prediction sites, `x.pred`, is:

x.pred

```
.03333 .03333
.03333 .96667
.50000 .50000
.96667 .03333
.96667 .96667
```

The order of the columns in `x.pred` *must be exactly the same as that in* `lhs.21`. Here, the x_1 values are in the first column of `x.pred` and the x_2 values are in the second column.

The following job file is used to obtain the REML estimates of the correlation parameters and the REML-EBLUP at the five sites in `x.pred`.

PErK job for Example C.5

```
# Definition stage

CorrelationFamily = Matern
CorrelationType = 2 # Specifies Matern II class
RandomError = No
Tries = 5
LogLikelihoodTolerance = 1.e-2
SimplexTolerance = 1.e-2
RandomNumberSeed = 400
CorrelationEstimation = REML

# Input stage

X < lhs.21
Y < y.21
XPred < x.pred
RegressionModel < reg.constant

# Output stage

Summary > C.5.summary
RegressionModel > C.5.beta
StochasticProcessModel > C.5.corpar
Predictions > C.5.blup
```

The summary information from the fit produced by PErK is in `C.5.summary`:

C.5.summary

```
Number of Design Sites = 21
Correlation Family = Matern II
Random Error = No
Stochastic Process Variance = 60569716.46114
Restricted Log Likelihood = -55.00658
Number of Restricted Log Likelihood Evaluations = 1144

Condition Number of Cholesky Square Root
of Correlation Matrix = 34252.65116

Number of Prediction Sites = 5
```

The estimate of the regression mean is given in `C.5.beta`.

C.5.beta

```
The estimates of the linear model parameters are:

Parameter Beta

1         8550.91207
```

The estimates of the correlation parameters are given in `C.5.corpar`:

C.5.corpar

```
Correlation Family = Matern II

REML estimates of the correlation parameters are:

Case Range Smoothness

1         2.60844         3.25927
2         9.89366         50.00000
```

The values of the EBLUP and its estimated standard error at the prediction sites in `x.pred` are listed in `C.5.blup`.

C.5.blup

```
The Predictor and its Standard Error are:

Case Prediction Standard Error

1          239.645996          1.6740799
2          11.228855          2.0946561
3          23.966300          0.4405656
4          4.320353          1.4408522
5          143.739681          1.7415620
```

The cases in the file `C.5.blup` are listed in the same order as the prediction sites in `x.pred`. ■

Example C.6 We modify Example C.4 in two ways. Recall that

$$\text{ERMSPE}(\widehat{y}) = \sqrt{\frac{1}{n}\sum_{i=1}^{n}(y(\boldsymbol{x}_i) - \widehat{y}(\boldsymbol{x}_i))^2}$$

denotes the *empirical root mean squared prediction error* (ERMSPE) of $\widehat{y}(\cdot)$ at the input points $\boldsymbol{x}_1, \ldots, \boldsymbol{x}_n$. First, this example illustrates how `PErK` can be made to calculate the ERMSPE of the EBLUP given the true responses at the input sites listed in the file `x.pred`. Assume that the true responses corresponding to the input sites in `x.pred` are listed in the file `y.true`. Second, we show how to use previously computed correlation parameters to determine the EBLUP. The prediction accuracy is assessed by the *maximum absolute deviation* (MAD) of the predictions from the true values.

y.true

```
241.39814
9.60979
24.12996
4.58372
143.48640
```

Recall that the REML estimates of the correlation parameters determined in Example C.4 were saved in the file `C.4.unf.corpar`.

C.4.unf.corpar

```
0.18494678690057
7.5188141072541e-3
7.1497890547390e-5
0.
```

The following job file is used to carry out the objectives of this example:

PErK job for Example C.6

```
# Definition stage

CorrelationFamily = PowerExponential
RandomError = No
CorrelationEstimation = REML

# Input stage

X < lhs.21
Y < y.21
XPred < x.pred
YTrue < y.true
RegressionModel < reg.C.4
StochasticProcessModel < C.4.unf.corpar

# Output stage

Summary > C.6.summary
Predictions > C.6.blup
```

The summary information from the `PErK` fit is in `C.6.summary`:

C.6.summary

```
Number of Design Sites = 21
Correlation Family = Power Exponential
Random Error = No
Stochastic Process Variance = 193426331.35018
Restricted Log Likelihood = -40.80125

Condition Number of Cholesky Square Root
of Correlation Matrix = 12131.01914

Number of Prediction Sites = 5
Prediction RMSPE = 1.74260
Prediction MAD = 3.09354
Case = 4
```

The values of the PErK and its estimated standard error at the prediction sites in `x.pred` are located in the file `C.6.blup`.

C.6.blup

```
The Predictor and its Standard Error are:

Case Prediction Standard Error

1          240.258350          5.9730420
2          7.996035          5.1030068
3          24.714434          2.5866037
4          1.490179          5.6309975
5          144.656146          5.6553202
```

■

Example C.7 For the Branin function (C.3.1), assume that x_1 is a *control* variable and x_2 is an *environmental* variable. Suppose that x_2 has support on the ten points $\{x_{2,i}^{sp}\}_{i=1}^{10}$ in file `ev.mat`:

ev.mat

```
0.05
0.15
0.25
0.35
0.45
0.55
0.65
0.75
0.85
0.95
```

and the associated probabilities $\{w_i\}_{i=1}^{10}$ are located in the file `evwt.mat`.

evwt.mat

```
0.001953125
0.017578125
0.0703125
0.1640625
0.24609375
0.24609375
0.1640625
0.0703125
0.017578125
0.001953125
```

C.8.blup

```
The Predictor and its Standard Error are:

Case Prediction Standard Error
1          39.992512          0.8637710
2          23.043269          0.5376513
3          28.962950          0.3554728
4          61.444669          0.0856200
5          42.804862          0.2905761
```

Example C.9 Suppose the computer response is the original Branin function $y_B(\cdot, \cdot)$ with input variables x_1 and x_2 having domain $[-5, 10] \times [0, 15]$ and we wish to carry out the analysis of Example C.5. We list the lower and upper bounds for x_1 and x_2 in the file `ranges`. The following job file would be used to carry out the analysis; the `Ranges` command is added to the *input* stage. ■

PErK job for Example C.9

```
# Definition stage
CorrelationFamily = Matern
CorrelationType = 2
RandomError = No
Tries = 5
LogLikelihoodTolerance = 1.e-2
SimplexTolerance = 1.e-2
RandomNumberSeed = 400
CorrelationEstimation = REML

# Input stage
X < lhs-original-scale.21
Ranges < ranges
Y < y.21
XPred < x-original-scale.pred
RegressionModel < reg.constant

# Output stage
Summary > C.9.summary
RegressionModel > C.9.beta
StochasticProcessModel > C.9.corpar
Predictions > C.9.blup
```

ranges

```
-5 10
0 15
```

References

Abrahamsen, P. (1997). A review of Gaussian random fields and correlation functions. *Technical Report 917.* Norwegian Computing Center, Box 114, Blindern, N0314 Oslo, Norway.

Abramowitz, M. and Stegun, I. (1965). *Handbook of Mathematical Functions.* Dover, New York.

Abt, M., Welch, W. J. and Sacks, J. (1998). Design and analysis for modeling and predicting spatial contamination. *Technical Report 98-82.* National Institute of Statistical Sciences.

Adler, R. J. (1981). *The Geometry of Random Fields.* J. Wiley, New York.

Adler, R. J. (1990). *An Introduction to Continuity, Extrema, and Related Topics for General Gaussian Processes.* Institute of Mathematical Statistics, Hayward, California.

Aitchison, J. and Dunsmore, I. R. (1975). *Statistical Prediction Analysis.* University Press, Cambridge.

Allen, D. M. (1974). The relationship between variable selection and data augmentation and a method for prediction. *Technometrics* **16**, 125–127.

An, J. and Owen, A. B. (1999). Quasi-regression. *Technical Report TR1999-23.* Department of Statistics, Stanford University.

Aslett, R., Buck, R. J., Duvall, S. G., Sacks, J. and Welch, W. J. (1998). Circuit optimization via sequential computer experiments: design of an output buffer. *Applied Statistics* **47**, 31–48.

Atkinson, A. C. and Donev, A. N. (1992). *Optimum experimental designs.* Oxford University Press.

Barry, R. P. and Ver Hoef, J. M. (1996). Blackbox kriging: spatial prediction without specifying variogram models. *Journal of Agricultural, Biological, and Environmental Statistics* **1**, 297–322.

Bartel, D. L., Burstein, A. H., Toda, M. D. and Edwards, D. L. (1985). The effect of conformity and plastic thickness on contact stresses in metal-backed plastic implants. *Journal of Biomechanical Engineering* **107**, 193–199.

Bates, R. A., Buck, R. J., Riccomagno, E. and Wynn, H. P. (1996). Experimental design and observation for large systems. *Journal of the Royal Statistical Society B* **58**, 77–94.

Bayarri, M. J., Berger, J. O., Higdon, D., Kennedy, M., Kottas, A., Paulo, R., Sacks, J., Cafeo, J., Cavendish, J., Lin, C. and Tu, J. (2002). A framework for validation of computer models. *Technical Report 128.* National Institute of Statistical Sciences.

Berger, J. O., De Oliveira, V. and Sansó, B. (2001). Objective Bayesian analysis of Spatially correlated data. *Journal of the American Statistical Association* **96**, 1361–1374.

Berk, R., Bickel, P., Campbell, K., Fovell, R., Keller-McNulty, S., Kelly, E., Linn, R., Park, B., Perelson, A., Rouphail, N., Sacks, J. and Schoenberg, F. (2002). Workshop on statistical approaches for the evaluation of complex computer models. *Statistical Science* **17**(2), 173–192.

Bernardo, M. C., Buck, R. J., Liu, L., Nazaret, W. A., Sacks, J. and Welch, W. J. (1992). Integrated circuit design optimization using a sequential strategy. *IEEE Transactions on Computer-Aided Design* **11**, 361–372.

Birk, D. M. (1997). *An Introduction to Mathematical Fire Modeling.* Technomic Publishing, Lancaster, PA.

Bochner, S. (1933). Monotone funktionen Stieltjessche integrale und harmonische analyse. *Mathematical Annals* **108**, 378.

Bochner, S. (1955). *Harmonic Analysis and the Theory of Probability.* University of California Press, Berkeley.

Booker, A. J. (1996). Case Studies in Design and Analysis of Computer Experiments. *ASA Proceedings of the Section on Physical and Engineering Sciences* (), pp. 244–248, American Statistical Association (Alexandria, VA).

Booker, A. J., Dennis, J. E., Frank, P. D., Serafini, D. B. and Torczon, V. (1997). Optimization Using surrogate objectives on a helicopter test example. *Technical Report SSGTECH-97-027.* Boeing Technical Report.

Booker, A. J., Dennis, J. E., Frank, P. D., Serafini, D. B., Torczon, V. and Trosset, M. W. (1999). A rigorous framework for optimization of expensive functions by surrogates. *Structural Optimization* **17**, 1–13.

Box, G. E. and Draper, N. R. (1987). *Empirical model-building and response surfaces.* John Wiley & Sons, New York.

Box, G. E. and Jones, S. (1992). Split-plot Designs for Robust Product Experimentation. *Journal of Applied Statistics* **19**, 3–26.

Bratley, P., Fox, B. L. and Niederreiter, H. (1994). Algorithm 738: Programs to Generate Niederreiter's Low-Discrepancy Sequences. *ACM Transactions on Mathematical Software* **20**, 494–495.

Buonaccorsi, J. P. and Iyer, H. K. (1986). Optimal designs for ratios of linear combinations in the general linear model. *Journal of Statistical Planning and Inference* **13**, 345–356.

Bursztyn, D. and Steinberg, D. (2001). Rotation designs for experiments in high bias situations. *Journal of Statistical Planning and Inference* **97**, 399–414.

Bursztyn, D. and Steinberg, D. M. (2002). Orthogonal first-order designs with higher order projectivity. *Journal of Applied Stochastic Models in Business and Industry* **18**, 197–206.

Bursztyn, D. and Steinberg, D. M. (2003). Comparison of designs for computer experiments. *Technical report.* Department of Statistics and Operations Research, Tel Aviv University.

Butler, N. A. (2001). Optimal and Orthogonal Latin Hypercube Designs for Computer Experiments. *Biometrika* **88**, 847–857.

Chang, P. B. (1998). *Robust design and analysis of femoral components for total hip arthroplasty.* PhD thesis. Sibley School of Mechanical and Aerospace Engineering, Cornell University. Ithaca, NY USA.

Chang, P. B., Williams, B. J., Bawa Bhalla, K. S., Belknap, T. W., Santner, T. J., Notz, W. I. and Bartel, D. L. (2001). Robust design and analysis of total joint replacements: Finite element model experiments with environmental variables. *Journal of Biomechanical Engineering* **123**, 239–246.

Chang, P. B., Williams, B. J., Notz, W. I., Santner, T. J. and Bartel, D. L. (1999). Robust optimization of total joint replacements incorporating environmental variables. *Journal of Biomechanical Engineering* **121**, 304–310.

Chiles, J. and Delfiner, P. (1999). *Geostatistcs: Modeling Spatial Uncertainty.* Wiley, New York.

Cleveland, W. S. (1993). *Visualizing Data.* Hobart Press, Summit, NJ.

Cooper, L. Y. (1980). Estimating safe available egress time from fires. *Technical Report 80-2172.* National Bureau of Standards, Washington D.C.

Cooper, L. Y. (1997). VENTCF2: An algorithm and associated FORTRAN 77 subroutine for calculating flow through a horizontal ceiling/floor vent in a zone-type compartmental fire model. *Fire Safety Journal* **28**, 253–287.

Cooper, L. Y. and Stroup, D. W. (1985). ASET–a computer program for calculating available safe egress time. *Fire Safety Journal* **9**, 29–45.

Cox, D. D., Park, J. S. and Singer, C. E. (1996). A statistical method for tuning a computer code to a database. *Technical Report 96-3.* Department of Statistics, Rice University.

Craig, P. C., Goldstein, M., Rougier, J. C. and Seheult, A. H. (2001). Bayesian Forecasting for Complex Systems using Computer Simulators. *Journal of the American Statistical Association* **96**, 717–729.

Craig, P. S., Goldstein, M., Seheult, A. H. and Smith, J. A. (1996). Bayes linear strategies for history matching of hydrocarbon reservoirs. In *Bayesian Statistics*, Vol. 5 (J. M. Bernardo, J. O. Berger, A. P. Dawid and A. F. M. Smith (eds)), pp. 69–95, Oxford University Press.

Craig, P. S., Goldstein, M., Seheult, A. H. and Smith, J. A. (1997). Pressure matching for hydrocarbon reservoirs: A case study in the use of Bayes linear strategies for large computer experiments (with discussion). In *Case Studies in Bayesian Statistics*, Vol. 3 (C. Gatsonis, J. Hodges, R. E. Kass, R. McCulloch, P. Rossi and N. Singpurwalla (eds)), pp. 36–93, Oxford University Press.

Craig, P. S., Goldstein, M., Seheult, A. H. and Smith, J. A. (1998). Constructing partial prior specifications for models of complex physical systems. *The Statistician* **47**, 37–53.

Cramér, H. and Leadbetter, M. R. (1967). *Stationary and Related Stochastic Processes.* J. Wiley, New York.

Cressie, N. A. (1993). *Statistics for Spatial Data.* J. Wiley, New York.

Crick, M. J., Hofer, E., Jones, J. A. and Haywood, S. M. (1988). Uncertainty analysis of the foodchain and atmospheric dispersion modules of MARC. *Technical Report NRPBR184.* National Radiological Protection Board.

Currin, C., Mitchell, T. J., Morris, M. D. and Ylvisaker, D. (1991). Bayesian prediction of deterministic functions, with applications to the design and analysis of computer experiments. *Journal of the American Statistical Association* **86**, 953–963.

Dandekar, R. and Kirkendall, N. (1993). Latin hypercube sampling for sensitivity and uncertainty analysis. *ASA Proceedings of the Section on Physical and Engineering Sciences* (), pp. 26–31, American Statistical Association.

De Oliveira, V. and Ecker, M. D. (2002). Bayesian hot spot detection in the presence of a spatial trend: application to total notrogen concentration in Chesapeake Bay. *Environmetrics* **13**, 85–101.

De Oliveira, V., Kedem, B. and Short, D. A. (1997). Bayesian prediction of transformed Gaussian random fields. *Journal of the American Statistical Association* **92**, 1422–1433.

Dean, A. M. and Voss, D. (1999). *Design and Analysis of Experiments.* Spring Verlag, New York.

Dimnaku, A., Kincaidy, R. and Trosset, M. W. (2002). Approximate solutions of continuous dispersion problems. *Technical report.* Department of Mathematics, College of William & Mary.

Dixon, L. C. W. and Szego, G. P. (1978). The global optimisation problem: an introduction. In *Towards Global Optimisation*, Vol. 2 (L. C. W. Dixon and G. P. Szego (eds)), pp. 1–15, North Holland, Amsterdam.

Doob, J. L. (1953). *Stochastic Processes.* J. Wiley, New York.

Draper, D. (1995). Assessment and propagation of model uncertainty. *Journal of the Royal Statistical Society B* **57**, 45–97.

Draper, D., Pereira, A., Prado, P., Saltelli, A., Cheal, R., Eguilior, S., Mendes, B. and Tarantola, S. (1999). Scenario and parametric uncertalnty in GESAMAC: A methodological study in nuclear waste displosal risk assessment. *Computer Physics Communications* **117**, 142–155.

Draper, N. R. and Smith, H. (1981). *Applied Regression Analysis, 2nd Ed.* J. Wiley, New York.

Duckworth, W. M. (2000). Some Binary Maximin Distance Designs. *Journal of Statistical Planning and Inference* **88**(1), 149–170.

Dunnett, C. W. (1989). Multivariate normal probability integrals with product correlation structure. Correction: **42**, 709. *Applied Statistics* **38**, 564–579.

Easterling, R. G. (1999). A framework for model validation. *Technical Report SAND99-0301C.* Sandia National Laboratories.

Easterling, R. G. (2001). Measuring the Predictive Capability of Computational Models: Principles and Methods, Issues and Illustrations. *Technical Report SAND2001-0243.* Sandia National Laboratories.

Ecker, M. D. and Gelfand, A. E. (1997). Bayesian variogram modeling for an isotropic spatial process. *Journal of Agricultural, Biological, and Environmental Statistics* **2**, 347–369.

Fang, K.-T., Lin, D. K., Winker, P. and Zhang, Y. (2000). Uniform Design: Theory and Application. *Technometrics* **42**, 237–248.

Frenklach, M., Wang, H. and Rabinowitz, M. J. (1992). Optimization and analysis of large chemical kinetic mechanisms using the solution mapping method–combustion of methane. *Progress Energy Combustion Science* **18**, 47–73.

Fuller, W. A. and Hasza, D. P. (1981). Properties of predictors for autoregressive time series (Corr: V76, 1023–1023). *Journal of the American Statistical Association* **76**, 155–161.

Furnival, G. M. and Wilson Jr., R. W. (1974). Regression by leaps and bounds. *Technometrics* **16**, 499–511.

Galway, L. and Lucas, T. (1997). Pressure matching for hydrocarbon reservoirs. *Case Studies in Bayesian Statistics* **3**, 87–91.

Gibbs, M. N. (1997). *Bayesian Gaussian Processes for Regression and Classification.* PhD thesis. Cambridge University. Cambridge, UK.

Giglio, B., Bates, R. A. and Wynn, H. P. (2000). Gröbner bases strategies in regression. *Journal of Applied Statistics* **27(7)**, 923–938.

Gikhman, I. I. and Skorokhod, A. V. (1977). *Introduction Into the Theory of Random Processes (2nd Ed) (Russian)*. Science Publishing House [Izdatelstvo Nauka].

Gikhman, I. I. and Skorokhod, A. V. (1979). *The Theory of Stochastic Processes, I II (S. Kotz (Transl.))*. Springer-Verlag, New York.

Goldstein, M. (1999). Bayes Linear Analysis. *Encyclopedia of Statistical Sciences: Update* **3**, 29–34.

Goldstein, M. and Rougier, J. C. (2002). Calibrated Bayesian forecasting using large computer simulators. *Technical report.* Department of Mathematical Sciences, University of Durham.

Golub, G. H., Heath, M. and Wahba, G. (1979). Generalized Cross-validation As a Method for Choosing a Good Ridge Parameter. *Technometrics* **21**, 215–223.

Gough, W. A. and Welch, W. J. (1993). Parameter space of an ocean general circulation model using an isopycnal mixing parameterization. *Technical Report 93-03*. National Institute of Statistical Sciences.

Graves, T. L. and Mockus, A. (1998). Inferring change effort from configuration management databases. *Technical Report 98-84*. National Institute of Statistical Sciences.

Guttorp, P. and Sampson, P. D. (1994). Methods for estimating heterogeneous spatial covariance functions with environmental applications. In *Handbook of Statistics*, Vol. 12 (G. P. Patil and C. R. Rao (eds)), pp. 661–689, Elsevier Science B.V.

Guttorp, P., Meiring, W. and Sampson, P. D. (1994). A Space-Time analysis of ground-level ozone data. *Environmetrics* **5**, 241–254.

Handcock, M. S. (1991). On cascading latin hypercube designs and additive models for experiments. *Communications Statistics—Theory Methods* **20**, 417–439.

Handcock, M. S. and Stein, M. L. (1993). A Bayesian analysis of kriging. *Technometrics* **35**, 403–410.

Handcock, M. S. and Wallis, J. R. (1994). An approach to statistical spatial-temporal modeling of meterological fields. *Journal of the American Statistical Association* **89**, 368–390.

Harville, D. A. (1974). Bayesian inference for variance components using only error contrasts. *Biometrika* **61**, 383–385.

Harville, D. A. (1977). Maximum likelihood approaches to variance component estimation and to related problems (with discussion). *Journal of the American Statistical Association* **72**, 320–340.

Harville, D. A. (1985). Decomposition of prediction error. *Journal of the American Statistical Association* **80**, 132–138.

Harville, D. A. (1997). *Matrix algebra from a statistician's perspective.* Springer Verlag, New York.

Harville, D. A. and Jeske, D. R. (1992). Mean squared error of estimation or prediction under a general linear mode. *Journal of the American Statistical Association* **87**, 724–731.

Haslett, J. and Raftery, A. E. (1989). Space-time modelling with long-memory dependence: assessing Ireland's wind power resource (with discussion). *Applied Statistics* **38**, 1–50.

Hass, T. C. (1995). Local prediction of a spatio-temporal process with an application to wet sulfate deposition. *Journal of the American Statistical Association* **90**, 1189–1199.

Hastie, T., Tibshirani, R. and Friedman, J. (2001). *The Elements of Statistical Learning: Data Mining, Inference, and Prediction.* Springer Verlag, New York.

Haylock, R. G. and O'Hagan, A. (1996). On inference for outputs of computationally expensive algorithms with uncertainty on the inputs. In *Bayesian Statistics*, Vol. 5 (J. M. Bernardo, J. O. Berger, A. P. Dawid and A. F. M. Smith (eds)), pp. 629–637, Oxford University Press.

Hedayat, A., Sloane, N. and Stufken, J. (1999). *Orthogonal Arrays.* Springer Verlag, New York.

Helton, J. C. (1993). Uncertainty and sensitivity analysis techniques for use in performance assessment for radioactive waste disposal. *Reliability Engineering and System Safety* **42**, 327–367.

Helton, J. C., Garner, J. W., McCurley, R. D. and Rudeen, D. K. (1991). Sensitivity analysis techniques and results for performance assessment at the waste isolation pilot plant. *Technical Report SAND90-7103.* Sandia National Laboratories.

Heo, G., Schmuland, B. and Wiens, D. P. (2001). Restricted minimax robust designs for misspecified regression models. *The Canadian Journal of Statistics* **29**, 117–128.

Higdon, D. M. (1998a). Auxiliary Variable Methods for Markov Chain Monte Carlo With Applications. *Journal of the American Statistical Association* **93**, 585–595.

Higdon, D. M. (1998b). A Process-convolution Approach to Modelling Temperatures in the North Atlantic Ocean (Discussion: P191-192). *Environmental and Ecological Statistics* **5**, 173–190.

Higdon, D. M., Swall, J. and Kern, J. C. (1999). Non-stationary Spatial Modeling. In *Bayesian Statistics*, Vol. 6 (J. M. Bernardo, J. O. Berger, A. P. Dawid and A. F. M. Smith (eds)), pp. 761–768, Oxford University Press.

Hills, R. G. and Trucano, T. G. (1999). Statistical Validation of Engineering and Scientific Models: Background. *Technical Report SAND99-1256.* Sandia National Laboratories.

Hills, R. G. and Trucano, T. G. (2001). Statistical Validation of Engineering and Scientific Models: A Maximum Likelihood Based Metric. *Technical Report SAND2001-1783.* Sandia National Laboratories.

Homma, T. and Saltelli, A. (1996). Importance measures in global sensitivity analysis of model output. *Reliability Engineering and System Safety* **52**, 1–17.

Hoshino, N. and Takemura, A. (2000). On Reduction of Finite-sample Variance By Extended Latin Hypercube Sampling. *Bernoulli* **6**(6), 1035–1050.

Huber, P. J. (1981). *Robust Statistics.* J. Wiley, New York.

Iman, R. L. and Conover, W. J. (1980). Small-sample sensitivity analysis techniques for computer models, with an application to risk assessment (with discussion). *Communications in Statistics - Theory and Methods* **A9**, 1749–1874.

Iman, R. L. and Conover, W. J. (1982). A distribution-free approach to inducing rank correlation among input variables. *Communications in Statistics, Series B* **11**, 311–334.

Jeffreys, H. (1961). *Theory of Probability.* Oxford University Press, London.

John, J. A. (1987). *Cyclic Designs.* Chapman & Hall Ltd, New York.

John, P. W. M. (1980). *Incomplete Block Designs.* M. Dekker, Inc., New York.

Johnson, M. E., Moore, L. M. and Ylvisaker, D. (1990). Minimax and maximin distance designs. *Journal of Statistical Planning and Inference* **26**, 131–148.

Jones, D. R., Schonlau, M. and Welch, W. J. (1998). Efficient global optimization of expensive black–box functions. *Journal of Global Optimization* **13**, 455–492.

Journel, A. G. and Huijbregts, C. J. (1978). *Mining Geostatistics.* Academic Press, London.

Journel, A. G. and Huijbregts, C. J. (1979). *Mining Geostatistics.* Academic Press, New York.

Kackar, R. N. and Harville, D. A. (1984). Approximations for standard errors of estimators of fixed and random effects in mixed linear models. *Journal of the American Statistical Association* **87**, 853–862.

Kaiser, M. S., Hsu, N.-J., Cressie, N. A. and Lahiri, S. N. (1997). Inference for spatial processes using subsampling: a stimulation study. *Environmetrics* **8**, 485–502.

Kennedy, M. C. and O'Hagan, A. (2000). Predicting the output from a complex computer code when fast approximations are available. *Biometrika* **87**, 1–13.

Kennedy, M. C. and O'Hagan, A. (2001). Bayesian Calibration of Computer Models (with discussion). *Journal of the Royal Statistical Society B* **63**, 425–464.

Kern, J. C. and Higdon, D. M. (1999). A Distance Metric to Account for Edge Effects in Spatial Data Analysis. *ASA Proceedings of the Biometrics Section* (), pp. 49–52, American Statistical Association (Alexandria, VA).

Kimeldorf, G. S. and Wahba, G. (1970). A correspondence between Bayesian estimation on stochastic processes and smoothing by splines. *Annals of Mathematical Statistics* **41**, 495–502.

Kleijnen, J. and van Beers, W. (2003). Application-driven Sequential Designs for Simulation Experiments: Kriging Metamodeling. *Technical Report WP 11.* Department of Information Systems and Management and Center for Economic Research, Tilburg University.

Kleijnen, J., Sanchez, S. M., Lucas, T. and van Beers, W. (2003). A User's Guide to the Brave New World of Designing Simulation Experiments. *Technical Report WP 10.* Department of Information Systems and Management and Center for Economic Research, Tilburg University.

Koehler, J. R. and Owen, A. B. (1996). Computer experiments. In *Handbook of Statistics*, Vol. 13 (S. Ghosh and C. R. Rao (eds)), pp. 261–308, Elsevier Science B.V.

Kotzar, G. M., Davy, D. T., Berilla, J. and Goldberg, V. M. (1995). Torsional Loads in the Early Postoperative Period Following Total Hip Replacement. *Journal of Orthopaedic Research* **13**, 945–955.

Kozintsev, B. (1999). *Computations With Gaussian Random Fields.* PhD thesis. Department of Mathematics and Institute for Systems Research, University of Maryland. College Park, MD USA.

Kozintsev, B. and Kedem, B. (2000). Generation of 'similar' images from a given discrete image. *Journal of Computational and Graphical Statistics* **9**, 286–302.

Kreyszig, E. (1999). *Advanced engineering mathematics.* John Wiley, New York.

Laslett, G. M. (1994). Kriging and Splines: An Empirical Comparison of Their Predictive Performance in Some Applications (Disc: P 401-409). *Journal of the American Statistical Association* **89**, 391–400.

Lehman, J. (2002). *Sequential Design of Computer Experiments for Robust Parameter Design.* PhD thesis. Department of Statistics, Ohio State University. Columbus, OH USA.

Lehman, J., Santner, T. J. and Notz, W. I. (2003). Robust parameter design for computer experiments. *Technical Report 708.* Department of Statistics, The Ohio State University.

Lempert, R., Schlensinger, M. E., Bankes, S. and Andronova, N. (2000). The impacts of climate variability on near-term policy choices and the value of information. *Climate Change* **45**, 129–161.

Lim, Y. B., Sacks, J., Studden, W. J. and Welch, W. J. (2002). Design and analysis of computer experiments when the output is highly correlated over the input space. *The Canadian Journal of Statistics* **30**(1), 109–126.

Lindley, D. V. (1956). On a measure of information provided by an experiment. *Annals of Mathematical Statistics* **27**, 986–1005.

Loh, W.-L. (1996). On latin hypercube sampling. *The Annals of Statistics* **24**, 2058–2080.

Loh, W.-L. and Lam, T.-K. (2000). Estimating structered correlation matrices in smooth Gaussian random field models. *The Annals of Statistics* **28**, 880–904.

Lynn, R. R. (1997). Transport model for prediction of wildfire behavior. *Technical Report LA13334-T.* Los Alamos National Laboratory.

Matérn, B. (1960). *Spatial Variation.* PhD thesis. Meddelanden fran Statens Skogsforskningsinstitut. Vol. 49, Num. 5.

Matérn, B. (1986). *Spatial Variation (Second Edition).* Springer-Verlag, New York.

Matheron, G. (1963). Principles of geostatistics. *Economic Geology* **58**, 1246–1266.

McDonald, G. C. (1981). Confidence intervals for vehicle emission deterioration factors. *Technometrics* **23**, 239–242.

McKay, M. D., Beckman, R. J. and Conover, W. J. (1979). A comparison of three methods for selecting values of input variables in the analysis of output from a computer code. *Technometrics* **21**, 239–245.

Mitchell, T. J. (1974). An algorithm for the construction of "D-optimal" experimental designs. *Technometrics* **16**, 203–210.

Mitchell, T. J. and Morris, M. D. (1988). A Bayesian Approach to the Design and Analysis of Computational Experiments. *Computer Science and Statistics: Proceedings of the 20th Symposium on the Interface* (), pp. 49–51, American Statistical Association (Alexandria, VA).

Mitchell, T. J. and Scott, D. S. (1987). A computer program for the design of group testing experiments. *Communications in Statistics - Theory and Methods* **16**, 2943–2955.

Mitchell, T. J., Morris, M. D. and Ylvisaker, D. (1990). Existence of Smoothed Stationary Processes on An Interval. *Stochastic Processes and their Applications* **35**, 109–119.

Mitchell, T. J., Morris, M. D. and Ylvisaker, D. (1994). Asymptotically optimum experimental designs for prediction of deterministic functions given derivative information. *Journal of Statistical Planning and Inference* **41**, 377–389.

Mockus, A. (1998). Estimating dependencies from spatial averages. *Journal of Computational and Graphical Statistics* **7**, 501–513.

Mockus, A., Mockus, J. and Mockus, L. (1994). Bayesian approach adapting stochastic and heuristic methods of global and discrete optimization (STMA V38 0013). *Informatica (Vilnius)* **5**, 123–166.

Mockus, J. (1994). Application of Bayesian approach to numerical methods of global and stochastic optimization. *Journal of Global Optimization* **4**, 347–365.

Mockus, J., Eddy, W., Mockus, A., Mockus, L. and Reklaitis, G. (1997). *Bayesian Heuristic Approach to Discrete and Global Optimization: Algorithms, Visualization, Software, and Applications.* Kluwer Academic, New York.

Mockus, J., Tiešis, V. and Žilinskas, A. (1978). The application of Bayesian methods for seeking the extremum. In *Towards Global Optimisation*, Vol. 2 (L. C. W. Dixon and G. P. Szego (eds)), pp. 117–129, North Holland, Amsterdam.

Montgomery, P. and Truss, L. T. (2001). Combining a statistical design of experiments with formability simulations to predict the formability of pockets in sheet metal parts. *Society of Automotive Engineers.*

Morris, M. D. (1991). Factorial sampling plans for preliminary computational experiments. *Technometrics* **33**, 161–174.

Morris, M. D. and Mitchell, T. J. (1995). Exploratory designs for computational experiments. *Journal of Statistical Planning and Inference* **43**, 381–402.

Morris, M. D., Mitchell, T. J. and Ylvisaker, D. (1993). Bayesian design and analysis of computer experiments: use of derivatives in surface prediction. *Technometrics* **35**, 243–255.

Mrawira, D., Welch, W. J., Schonlau, M. and Haas, R. (1999). Sensitivity analysis of computer models: the world bank HDM-III model. *Journal of Transportation Engineering* **125**, 421–428.

Müller-Gronbach, T. (1996). Optimal designs for approximating the path of a stochastic process. *Journal of Statistical Planning and Inference* **49**, 371–385.

Naylor, J. C. and Smith, A. F. M. (1982). Applications of a method for the efficient computation of posterior distributions. *Applied Statistics* **31**, 214–225.

Neal, R. (1999). Regression and classification using Gaussian process priors. *Bayesian Statistics* **6**, 475–501.

Neal, R. M. (2003). Slice Sampling (with discussion). *Annals of Statistics* **31**, to appear.

Nelder, J. A. and Mead, R. (1965). A simplex method for function minimization. *Computer Journal* **7**, 308–313.

Niederreiter, H. (1992). *Random Number Generation and Quasi-Monte Carlo Methods.* SAIM, Philadelphia.

Oakley, J. E. (2002). Eliciting Gaussian Process Priors for Complex Computer Codes. *The Statistician* **51**, 81–97.

Oakley, J. E. and O'Hagan, A. (2002). Bayesian inference for the uncertainty distribution of computer model outputs. *Biometrika* **89** **(4)**, 769–784.

Odeh, R., Davenport, J. and Pearson, N. (eds) (1988). *Selected Tables in Mathematical Statistics (Volume 11).* American Mathematical Society.

O'Hagan, A. (1978). Curve fitting and optimal design for prediction. *Journal of the Royal Statistical Society B* **40**, 1–42.

O'Hagan, A. (1991). Bayes–Hermite quadrature. *Journal of Statistical Planning and Inference* **29**, 245–260.

O'Hagan, A. (1992). Some Bayesian numerical analysis. In *Bayesian Statistics*, Vol. 4 (J. M. Bernardo, J. O. Berger, A. P. Dawid and A. F. M. Smith (eds)), pp. 345–363, Oxford University Press.

O'Hagan, A. (1998). A Markov property for covariance structures. *Technical Report 98-13.* University of Nottingham, Nottingham, UK.

O'Hagan, A. and Haylock, R. G. (1997). Bayesian uncertainty analysis and radiological protection. In *Statistics for the Environment*, Vol. 3 (V. Barnett and K. F. Turkman (eds)), pp. 109–128, J. Wiley.

O'Hagan, A., Kennedy, M. C. and Oakley, J. E. (1999). Uncertainty analysis and other inference tools for complex computer codes. In *Bayesian Statistics*, Vol. 6 (J. M. Bernardo, J. O. Berger, A. P. Dawid and A. F. M. Smith (eds)), pp. 503–524, Oxford University Press.

Omre, H. (1987). Bayesian kriging-merging observations and qualified guesses in kriging. *Mathematical Geology* **19**, 25–39.

Owen, A. B. (1992a). A central limit theorem for latin hypercube sampling. *Journal of the Royal Statistical Society B* **54**, 541–551.

Owen, A. B. (1992b). Orthogonal arrays for computer experiments, integration and visualization. *Statistica Sinica* **2**, 439–452.

Owen, A. B. (1994). Controlling correlations in latin hypercube samples. *Journal of the American Statistical Association* **89**, 1517–1522.

Owen, A. B. (1995). Randomly permuted (t, m, s)-nets and (t, s) sequences. In *Monte Carlo and Quasi-Monte Carlo Methods in Scientific Computing* (H. Niederreiter and P. J.-S. Shiue (eds)), pp. 299–317, Springer-Verlag, New York.

Palmer, K. and Tsui, K.-L. (2001). A minimum bias latin hypercube design. *IEEE Transactions* **33**, 793–808.

Park, J. S. (1991). *Tuning Complex Computer Codes to Data and Optimal Designs.* PhD thesis. University of Illinois. Champaign/Urbana, IL USA.

Park, J. S. (1994). Optimal Latin-hypercube designs for computer experiments. *Journal of Statistical Planning and Inference* **39**, 95–111.

Park, J. S. (1999). Statistical estimation of the input parameters in complex simulation code. *Korean Journal of Applied Statistics* **12**, 335–345.

Park, J. S. and Hwang, H. S. (2000). Comparison and Distance Calculation Between Two Latin-hypercube Designs. *Korean Journal of Applied Statistics* **13**(2), 477–488.

Parzen, E. (1962). *Stochastic Processes.* Holden-Day, San Francisco.

Patterson, H. D. and Thompson, R. (1971). Recovery of interblock information when block sizes are unequal. *Biometrika* **58**, 545–554.

Patterson, H. D. and Thompson, R. (1974). Maximum likelihood estimation of components of variance. *Proceedings of the 8th International Biometric Conference* (), pp. 197–207, Biometric Society, Washington DC.

Pebesma, E. J. and Heuvelink, G. B. M. (1999). Latin Hypercube Sampling of Gaussian Random Fields. *Technometrics* **41**, 303–312.

Piepel, G. F. (1997). Survey of software with mixture experiment capabilities. *Journal of Quality Technology* **29**, 76–85.

Pilch, M., Trucano, T. G., Moya, J., Froehlich, G., Hodges, A. and Peercy, D. (2000). Guidelines for Sandia ASCI Verification and Validation Plans Content and Format: Version 2.0. *Technical Report SAND2000-3101.* Sandia National Laboratories.

Poole, D. and Raftery, A. E. (1998). Inference for deterministic simulation models: the Bayesian melding approach. *Technical Report 346.* Department of Statistics, University of Washington.

Prasad, N. G. N. and Rao, J. N. K. (1990). The estimation of the mean squared error of small-area estimators. *Journal of the American Statistical Association* **85**, 163–171.

Pukelsheim, F. (1993). *Optimal Design of Experiments.* J. Wiley, New York.

Raftery, A. E., Givens, G. H. and Zeh, J. E. (1995). Inference from a deterministic population dynamics model for bowhead whales (with discussion). *Journal of the American Statistical Association* **90**, 401–430.

Raghavan, N., Goel, P. and Ghosh, S. (1997). Classification of Mixtures of Spatial Point Processes. *PCmpSSt*, Vol. 29 (), pp. 112–116, American Statistical Association.

Raghavarao, D. (1971). *Constructions and Combinatorial Problems in Design of Experiments.* J. Wiley, New York.

Raghavarao, D. (1988). *Constructions and Combinatorial Problems in Design of Experiments.* Dover.

Reese, C. S., Wilson, A. G., Hamada, M., Martz, F. and Ryan, K. (2000). Integrated analysis of computer and physical experiments. *Technical Report LA-UR-00-2915.* Sandia Laboratories.

Rinnooy Kan, A. H. G. and Timmer, G. T. (1984). A stochastic approach to global optimization. In *Optimization 84: Proceedings of the SIAM Conference on Numerical Optimization* (P. T. Boggs, R. H. Byrd and R. B. Schnabel (eds)), pp. 245–262, SIAM, Philadelphia.

Ripley, B. D. (1981). *Spatial Statistics.* Wiley.

Roache, P. J. (1998). *Verification and Validation in Computational Science and Engineering.* Hermosa Publishers, Albuquerque.

Robert, C. P. and Casella, G. (1999). *Monte Carlo Statistical Methods.* Springer-Verlag.

Rodríguez-Iturbe, I. and Mejía, J. M. (1974). The design of rainfall networks in time and space. *Water Resources Research* **10**, 713–728.

Romano, D. and Vicario, G. (2001). Reliable Estimation in Computer Experiments on Finite-element Codes. *Quality Engineering* **14**(2), 195–204.

Sacks, J., Rouphail, N. M., Park, B. and Thakuriah, P. (2000). Statistically-based validation of computer simulation models in traffic operations and management. *Technical Report 112.* National Institute of Statistical Sciences.

Sacks, J., Schiller, S. B. and Welch, W. J. (1992). Design for computer experiments. *Technometrics* **31**, 41–47.

Sacks, J., Welch, W. J., Mitchell, T. J. and Wynn, H. P. (1989). Design and analysis of computer experiments. *Statistical Science* **4**, 409–423.

Sahama, A. R. and Diamond, N. T. (2001). Sample Size Considerations and Augmentation of Computer Experiments. *Journal of Statistical Computation and Simulation* **68**(4), 307–319.

Saltelli, A. (2002). Making best use of model evaluations to compute sensitivity indices. *Computer Physics Communications* **145 (2)**, 280–297.

Saltelli, A., Chan, K. and Scott, E. (2000). *Sensitivity Analysis.* John Wiley & Sons, Chichester.

Yaglom, A. M. (1986a). *Correlation theory of stationary and related random functions (Vol. 1)*. Springer-Verlag, New York.

Yaglom, A. M. (1986b). *Correlation theory of stationary and related random functions (Vol. II)*. Springer-Verlag, New York.

Yaglom, M. I. (1962). *Introduction to the Theory of Stationary Random Functions*. Dover, New York.

Ye, K. Q. (1998). Orthogonal column latin hypercubes and their application in computer experiments. *Journal of the American Statistical Association* **93**, 1430–1439.

Ying, Z.-L. (1991). Asymptotic Properties of a Maximum Likelihood Estimator With Data From a Gaussian Process. *Journal of Multivariate Analysis* **36**, 280–296.

Ying, Z.-L. (1993). Maximum likelihood estimation of parameters under a spatial sampling scheme. *The Annals of Statistics* **21**, 1567–1590.

Zhang, S. and Porostosky, J. (1998). SPeDE user's guide with examples. *Technical Report 648*. Department of Statistics, The Ohio State University.

Zhang, S., Notz, W. I., Santner, T. J. and Bartel, D. L. (1999). Empirical BLUPS for computer experiments based on spectral density estimation. *Technical Report 6??* Department of Statistics, The Ohio State University.

Zimmerman, D. L. and Cressie, N. A. (1992). Mean squared prediction error in the spatial linear model with estimated covariance parameters. *Annals of the Institute of Statistical Mathematics* **44**, 27–43.

Subject Index

Author Index

Springer Series in Statistics

(continued from p. ii)

Knottnerus: Sample Survey Theory: Some Pythagorean Perspectives.
Kolen/Brennan: Test Equating: Methods and Practices.
Kotz/Johnson (Eds.): Breakthroughs in Statistics Volume I.
Kotz/Johnson (Eds.): Breakthroughs in Statistics Volume II.
Kotz/Johnson (Eds.): Breakthroughs in Statistics Volume III.
Küchler/Sørensen: Exponential Families of Stochastic Processes.
Lahiri: Resampling Methods for Dependent Data.
Le Cam: Asymptotic Methods in Statistical Decision Theory.
Le Cam/Yang: Asymptotics in Statistics: Some Basic Concepts, 2nd edition.
Liu: Monte Carlo Strategies in Scientific Computing.
Longford: Models for Uncertainty in Educational Testing.
Manski: Partial Identification of Probability Distributions.
Mielke/Berry: Permutation Methods: A Distance Function Approach.
Pan/Fang: Growth Curve Models and Statistical Diagnostics.
Parzen/Tanabe/Kitagawa: Selected Papers of Hirotugu Akaike.
Politis/Romano/Wolf: Subsampling.
Ramsay/Silverman: Applied Functional Data Analysis: Methods and Case Studies.
Ramsay/Silverman: Functional Data Analysis.
Rao/Toutenburg: Linear Models: Least Squares and Alternatives.
Reinsel: Elements of Multivariate Time Series Analysis, 2nd edition.
Rosenbaum: Observational Studies, 2nd edition.
Rosenblatt: Gaussian and Non-Gaussian Linear Time Series and Random Fields.
Särndal/Swensson/Wretman: Model Assisted Survey Sampling.
Santner/Williams/Notz: The Design and Analysis of Computer Experiments.
Schervish: Theory of Statistics.
Shao/Tu: The Jackknife and Bootstrap.
Simonoff: Smoothing Methods in Statistics.
Singpurwalla and Wilson: Statistical Methods in Software Engineering: Reliability and Risk.
Small: The Statistical Theory of Shape.
Sprott: Statistical Inference in Science.
Stein: Interpolation of Spatial Data: Some Theory for Kriging.
Taniguchi/Kakizawa: Asymptotic Theory of Statistical Inference for Time Series.
Tanner: Tools for Statistical Inference: Methods for the Exploration of Posterior Distributions and Likelihood Functions, 3rd edition.
van der Laan: Unified Methods for Censored Longitudinal Data and Causality.
van der Vaart/Wellner: Weak Convergence and Empirical Processes: With Applications to Statistics.
Verbeke/Molenberghs: Linear Mixed Models for Longitudinal Data.
Weerahandi: Exact Statistical Methods for Data Analysis.
West/Harrison: Bayesian Forecasting and Dynamic Models, 2nd edition.

Analyzing Categorical Data
Jeffrey S. Simonoff

This book provides an introduction to the analysis of categorical data. The coverage is broad, using the loglinear Poisson regression model and logistic binomial regression models as the primary engines for methodology. Topics covered include count regression models, such as Poisson, negative binomial, zero-inflated, and zero-truncated models; loglinear models for two-dimensional and multidimensional contingency tables, including for square tables and tables with ordered categories; and regression models for two-category (binary) and multiple-category target variables, such as logistic and proportional odds models. All methods are illustrated with analyses of real data examples, many from recent subject area journal articles.
2003/504pp/hardcover/ISBN 0-387-00749-0
SPRINGER TEXTS IN STATISTICS

Partial Identification of Probability Distributions
Charles F. Manski

Sample data alone never suffice to draw conclusions about populations. Inference always requires assumptions about the population and sampling process. Statistical theory has revealed much about how strength of assumptions affects the precision of point estimates, but has had much less to say about how it affects the identification of population parameters. Indeed, it has been commonplace to think of identification as a binary event—a parameter is either identified or not—and to view point identification as a precondition for inference. Yet there is enormous scope for fruitful inference using data and assumptions that partially identify population parameters. This book explains why and shows how.
2003/196pp/hardcover/ISBN 0-387-00454-8
SPRINGER SERIES IN STATISTICS

Nonlinear Time Series
Jianqing Fan and Qiwei Yao

This book presents contemporary statistical methods and theory of nonlinear time series analysis. The principal focus is on nonparametric and semiparametric techniques developed in the last decade. It covers the techniques for modelling in state-space, in frequency-domain as well as in time-domain. To reflect the integration of parametric and nonparametric methods in analyzing time series data, the book also presents an up-to-date exposure of some parametric nonlinear models, including ARCH/GARCH models and threshold models. A compact view on linear ARMA models is also provided. Data arising in real applications are used throughout to show how nonparametric approaches may help to reveal local structure in high-dimensional data.
2003/576pp/hardcover/ISBN 0-387-95170-9
SPRINGER SERIES IN STATISTICS